CAMBRIDGE LIBRARY COLLECTION

Books of enduring scholarly value

Philosophy

This series contains both philosophical texts and critical essays about philosophy, concentrating especially on works originally published in the eighteenth and nineteenth centuries. It covers a broad range of topics including ethics, logic, metaphysics, aesthetics, utilitarianism, positivism, scientific method and political thought. It also includes biographies and accounts of the history of philosophy, as well as collections of papers by leading figures. In addition to this series, primary texts by ancient philosophers, and works with particular relevance to philosophy of science, politics or theology, may be found elsewhere in the Cambridge Library Collection.

Collected Essays

Known as 'Darwin's Bulldog', the biologist Thomas Henry Huxley (1825–95) was a tireless supporter of the evolutionary theories of his friend Charles Darwin. Huxley also made his own significant scientific contributions, and he was influential in the development of science education despite having had only two years of formal schooling. He established his scientific reputation through experiments on aquatic life carried out during a voyage to Australia while working as an assistant surgeon in the Royal Navy; ultimately he became President of the Royal Society (1883–5). Throughout his life Huxley struggled with issues of faith, and he coined the term 'agnostic' to describe his beliefs. This nine-volume collection of Huxley's essays, which he edited and published in 1893–4, demonstrates the wide range of his intellectual interests. Volume 8 contains public lectures given by Huxley, on themes as diverse as yeast, lobsters and palaeontology.

T0188085

Cambridge University Press has long been a pioneer in the reissuing of out-of-print titles from its own backlist, producing digital reprints of books that are still sought after by scholars and students but could not be reprinted economically using traditional technology. The Cambridge Library Collection extends this activity to a wider range of books which are still of importance to researchers and professionals, either for the source material they contain, or as landmarks in the history of their academic discipline.

Drawing from the world-renowned collections in the Cambridge University Library, and guided by the advice of experts in each subject area, Cambridge University Press is using state-of-the-art scanning machines in its own Printing House to capture the content of each book selected for inclusion. The files are processed to give a consistently clear, crisp image, and the books finished to the high quality standard for which the Press is recognised around the world. The latest print-on-demand technology ensures that the books will remain available indefinitely, and that orders for single or multiple copies can quickly be supplied.

The Cambridge Library Collection will bring back to life books of enduring scholarly value (including out-of-copyright works originally issued by other publishers) across a wide range of disciplines in the humanities and social sciences and in science and technology.

Collected Essays

VOLUME 8: DISCOURSES:
BIOLOGICAL AND GEOLOGICAL

THOMAS HENRY HUXLEY

CAMBRIDGE
UNIVERSITY PRESS

CAMBRIDGE UNIVERSITY PRESS

Cambridge, New York, Melbourne, Madrid, Cape Town,
Singapore, São Paolo, Delhi, Tokyo, Mexico City

Published in the United States of America by Cambridge University Press, New York

www.cambridge.org
Information on this title: www.cambridge.org/9781108040587

© in this compilation Cambridge University Press 2011

This edition first published 1894
This digitally printed version 2011

ISBN 978-1-108-04058-7 Paperback

This book reproduces the text of the original edition. The content and language reflect
the beliefs, practices and terminology of their time, and have not been updated.

Cambridge University Press wishes to make clear that the book, unless originally published
by Cambridge, is not being republished by, in association or collaboration with, or
with the endorsement or approval of, the original publisher or its successors in title.

COLLECTED ESSAYS

By T. H. HUXLEY

VOLUME VIII

DISCOURSES:

BIOLOGICAL & GEOLOGICAL

ESSAYS

BY

THOMAS H. HUXLEY

𝕷𝖔𝖓𝖉𝖔𝖓
MACMILLAN AND CO.
1894

RICHARD CLAY AND SONS, LIMITED,
LONDON AND BUNGAY.

PREFACE

THE contents of the present volume, with three
exceptions, are either popular lectures, or addresses
delivered to scientific bodies with which I have
been officially connected.   I am not sure which
gave me the more trouble.   For I have not been
one of those fortunate persons who are able to
regard a popular lecture as a mere *hors d'œuvre*,
unworthy of being ranked among the serious efforts
of a philosopher; and who keep their fame as
scientific hierophants unsullied by attempts—at
least of the successful sort—to be understanded
of the people.

On the contrary, I found that the task of
putting the truths learned in the field, the
laboratory and the museum, into language which,
without bating a jot of scientific accuracy shall be
generally intelligible, taxed such scientific and
literary faculty as I possessed to the uttermost;
indeed my experience has furnished me with no
better corrective of the tendency to scholastic
pedantry which besets all those who are absorbed

in pursuits remote from the common ways of men, and become habituated to think and speak in the technical dialect of their own little world, as if there were no other.

If the popular lecture thus, as I believe, finds one moiety of its justification in the self-discipline of the lecturer, it surely finds the other half in its effect on the auditory. For though various sadly comical experiences of the results of my own efforts have led me to entertain a very moderate estimate of the purely intellectual value of lectures; though I venture to doubt if more than one in ten of an average audience carries away an accurate notion of what the speaker has been driving at; yet is that not equally true of the oratory of the hustings, of the House of Commons, and even of the pulpit?

Yet the children of this world are wise in their generation; and both the politician and the priest are justified by results. The living voice has an influence over human action altogether indepen-dent of the intellectual worth of that which it utters. Many years ago, I was a guest at a great City dinner. A famous orator, endowed with a voice of rare flexibility and power; a born actor, ranging with ease through every part, from refined comedy to tragic unction, was called upon to reply to a toast. The orator was a very busy man, a charming conversationalist and by no means despised a good dinner; and, I imagine, rose with-

out having given a thought to what he was going
to say. The rhythmic roll of sound was admirable,
the gestures perfect, the earnestness impressive;
nothing was lacking save sense and, occasionally,
grammar. When the speaker sat down the
applause was terrific and one of my neighbours
was especially enthusiastic. So when he had
quieted down, I asked him what the orator had
said. And he could not tell me.

That sagacious person John Wesley, is reported
to have replied to some one who questioned the
propriety of his adaptation of sacred words to
extremely secular airs, that he did not see why the
Devil should be left in possession of all the best
tunes. And I do not see why science should not
turn to account the peculiarities of human nature
thus exploited by other agencies : all the more
because science, by the nature of its being, cannot desire to stir the passions, or profit by the
weaknesses, of human nature. The most zealous
of popular lecturers can aim at nothing more
than the awakening of a sympathy for abstract
truth, in those who do not really follow his arguments ; and of a desire to know more and better in
the few who do.

At the same time it must be admitted that the
popularization of science, whether by lecture or
essay, has its drawbacks. Success in this department has its perils for those who succeed. The
"people who fail" take their revenge, as we have

recently had occasion to observe, by ignoring all the rest of a man's work and glibly labelling him a mere popularizer. If the falsehood were not too glaring, they would say the same of Faraday and Helmholtz and Kelvin.

On the other hand, of the affliction caused by persons who think that what they have picked up from popular exposition qualifies them for discussing the great problems of science, it may be said, as the Radical toast said of the power of the Crown in bygone days, that it " has increased, is increasing, and ought to be diminished." The oddities of " English as she is spoke " might be abundantly paralleled by those of " Science as she is misunderstood " in the sermon, the novel, and the leading article; and a collection of the grotesque travesties of scientific conceptions, in the shape of essays on such trifles as " the Nature of Life " and the " Origin of All Things," which reach me, from time to time, might well be bound up with them.

The tenth essay in this volume unfortunately brought me, I will not say into collision, but into a position of critical remonstrance with regard to some charges of physical heterodoxy, brought by my distinguished friend Lord Kelvin, against British Geology. As President of the Geological Society of London at that time (1869), I thought I might venture to plead that we were not such heretics as we seemed to be; and that, even if we were, recantation would not affect the question of evolution.

I am glad to see that Lord Kelvin has just reprinted his reply to my plea,[1] and I refer the reader to it. I shall not presume to question anything, that on such ripe consideration, Lord Kelvin has to say upon the physical problems involved. But I may remark that no one can have asserted more strongly than I have done, the necessity of looking to physics and mathematics, for help in regard to the earliest history of the globe. (See pp. 108 and 109 of this volume.)

And I take the opportunity of repeating the opinion, that, whether what we call geological time has the lower limit assigned to it by Lord Kelvin, or the higher assumed by other philosophers; whether the germs of all living things have originated in the globe itself, or whether they have been imported on, or in, meteorites from without, the problem of the origin of those successive Faunæ and Floræ of the earth, the existence of which is fully demonstrated by paleontology remains exactly where it was.

For I think it will be admitted, that the germs brought to us by meteorites, if any, were not ova of elephants, nor of crocodiles; not cocoa-nuts nor acorns; not even eggs of shell-fish and corals; but only those of the lowest forms of animal and vegetable life. Therefore, since it is proved that,

[1] *Popular Lectures and Addresses.* II. Macmillan and Co. 1894.

from a very remote epoch of geological time, the
earth has been peopled by a continual succession
of the higher forms of animals and plants, these
either must have been created, or they have arisen
by evolution.   And in respect of certain groups of
animals, the well-established facts of paleontology
leave no rational doubt that they arose by the
latter method.

In the second place, there are no data whatever, which justify the biologist in assigning
any, even approximately definite, period of time,
either long or short, to the evolution of one
species from another by the process of variation
and selection.   In the ninth of the following
essays, I have taken pains to prove that the change
of animals has gone on at very different rates in
different groups of living beings; that some types
have persisted with little change from the paleo-
zoic epoch till now, while others have changed
rapidly within the limits of an epoch.   In 1862
(see below p. 303, 304) in 1863 (vol. II., p. 461)
and again in 1864 (*ibid.*, p. 89—91) I argued, not
as a matter of speculation, but, from paleonto-
logical facts, the bearing of which I believe, up to
that time, had not been shown, that any adequate hypothesis of the causes of evolution must
be consistent with progression, stationariness and
retrogression, of the same type at different epochs;
of different types in the same epoch; and that
Darwin's hypothesis fulfilled these conditions.

According to that hypothesis, two factors are at work, variation and selection. Next to nothing is known of the causes of the former process; nothing whatever of the time required for the production of a certain amount of deviation from the existing type. And, as respects selection, which operates by extinguishing all but a small minority of variations, we have not the slightest means of estimating the rapidity with which it does its work. All that we are justified in saying is that the rate at which it takes place may vary almost indefinitely. If the famous paint-root of Florida, which kills white pigs but not black ones, were abundant and certain in its action, black pigs might be substituted for white in the course of two or three years. If, on the other hand, it was rare and uncertain in action, the white pigs might linger on for centuries.

T. H. HUXLEY.

HODESLEA, EASTBOURNE,
    April, 1894.

CONTENTS

X

XI

I

ON A PIECE OF CHALK

[1868]

IF a well were sunk at our feet in the midst of
the city of Norwich, the diggers would very soon
find themselves at work in that white substance
almost too soft to be called rock, with which we
are all familiar as " chalk."

Not only here, but over the whole county of
Norfolk, the well-sinker might carry his shaft
down many hundred feet without coming to the
end of the chalk; and, on the sea-coast, where
the waves have pared away the face of the land
which breasts them, the scarped faces of the high
cliffs are often wholly formed of the same material.
Northward, the chalk may be followed as far as
Yorkshire; on the south coast it appears abruptly
in the picturesque western bays of Dorset, and
breaks into the Needles of the Isle of Wight;
while on the shores of Kent it supplies that long

line of white cliffs to which England owes her
name of Albion.

Were the thin soil which covers it all washed
away, a curved band of white chalk, here broader,
and there narrower, might be followed diagonally
across England from Lulworth in Dorset, to Flam-
borough Head in Yorkshire—a distance of over
280 miles as the crow flies. From this band to
the North Sea, on the east, and the Channel, on
the south, the chalk is largely hidden by other
deposits; but, except in the Weald of Kent and
Sussex, it enters into the very foundation of all
the south-eastern counties.

Attaining, as it does in some places, a thickness
of more than a thousand feet, the English chalk
must be admitted to be a mass of considerable
magnitude. Nevertheless, it covers but an insignificant portion of the whole area occupied by the
chalk formation of the globe, much of which has
the same general characters as ours, and is found
in detached patches, some less, and others more
extensive, than the English. Chalk occurs in
north-west Ireland; it stretches over a large part
of France,—the chalk which underlies Paris being,
in fact, a continuation of that of the London basin;
it runs through Denmark and Central Europe, and
extends southward to North Africa; while eastward, it appears in the Crimea and in Syria, and
may be traced as far as the shores of the Sea of
Aral, in Central Asia. If all the points at which

true chalk occurs were circumscribed, they would lie within an irregular oval about 3,000 miles in long diameter—the area of which would be as great as that of Europe, and would many times exceed that of the largest existing inland sea— the Mediterranean.

Thus the chalk is no unimportant element in the masonry of the earth's crust, and it impresses a peculiar stamp, varying with the conditions to which it is exposed, on the scenery of the districts in which it occurs. The undulating downs and rounded coombs, covered with sweet-grassed turf, of our inland chalk country, have a peacefully domestic and mutton-suggesting prettiness, but can hardly be called either grand or beautiful. But on our southern coasts, the wall-sided cliffs, many hundred feet high, with vast needles and pinnacles standing out in the sea, sharp and solitary enough to serve as perches for the wary cormorant, confer a wonderful beauty and grandeur upon the chalk headlands. And, in the East, chalk has its share in the formation of some of the most venerable of mountain ranges, such as the Lebanon.

What is this wide-spread component of the surface of the earth? and whence did it come?

You may think this no very hopeful inquiry. You may not unnaturally suppose that the attempt to solve such problems as these can lead to no result, save that of entangling the inquirer in vague speculations, incapable of refutation and of verification. If such were really the case, I should have selected some other subject than a " piece of chalk " for my discourse. But, in truth, after much deliberation, I have been unable to think of any topic which would so well enable me to lead you to see how solid is the foundation upon which some of the most startling conclusions of physical science rest.

A great chapter of the history of the world is written in the chalk. Few passages in the history of man can be supported by such an overwhelming mass of direct and indirect evidence as that which testifies to the truth of the fragment of the history of the globe, which I hope to enable you to read, with your own eyes, to-night. Let me add, that few chapters of human history have a more profound significance for ourselves. I weigh my words well when I assert, that the man who should know the true history of the bit of chalk which every carpenter carries about in his breeches-pocket, though ignorant of all other history, is likely, if he will think his knowledge out to its ultimate results, to have a truer, and therefore a better, conception of this wonderful universe, and of man's relation to it, than the most learned student who is deep-read in the records of humanity and ignorant of those of Nature.

The language of the chalk is not hard to learn, not nearly so hard as Latin, if you only want to get at the broad features of the story it has to tell; and I propose that we now set to work to spell that story out together.

We all know that if we "burn" chalk the result is quicklime. Chalk, in fact, is a compound of carbonic acid gas, and lime, and when you make it very hot the carbonic acid flies away and the lime is left. By this method of procedure we see the lime, but we do not see the carbonic acid. If, on the other hand, you were to powder a little chalk and drop it into a good deal of strong vinegar, there would be a great bubbling and fizzing, and, finally, a clear liquid, in which no sign of chalk would appear. Here you see the carbonic acid in the bubbles; the lime, dissolved in the vinegar, vanishes from sight. There are a great many other ways of showing that chalk is essentially nothing but carbonic acid and quicklime. Chemists enunciate the result of all the experiments which prove this, by stating that chalk is almost wholly composed of "carbonate of lime."

It is desirable for us to start from the knowledge of this fact, though it may not seem to help us very far towards what we seek. For carbonate of lime is a widely-spread substance, and is met with under very various conditions. All sorts of limestones are composed of more or less pure carbonate of lime. The crust which is often deposited by waters which have drained through limestone rocks, in the form of what are called stalagmites and stalactites, is carbonate of lime. Or, to take a more familiar example, the fur on the inside of a tea-kettle is carbonate of lime; and, for anything chemistry tells us to the contrary, the chalk might be a kind of gigantic fur upon the bottom of the earth-kettle, which is kept pretty hot below.

Let us try another method of making the chalk tell us its own history. To the unassisted eye chalk looks simply like a very loose and open kind of stone. But it is possible to grind a slice of chalk down so thin that you can see through it—until it is thin enough, in fact, to be examined with any magnifying power that may be thought desirable. A thin slice of the fur of a kettle might be made in the same way. If it were examined microscopically, it would show itself to be a more or less distinctly laminated mineral substance, and nothing more.

But the slice of chalk presents a totally different appearance when placed under the microscope. The general mass of it is made up of very minute granules; but, imbedded in this matrix, are innumerable bodies, some smaller and some larger, but, on a rough average, not more than a hundredth of an inch in diameter, having a well-defined shape and structure. A cubic inch of some specimens of chalk may contain hundreds of thousands of these bodies, compacted together with incalculable millions of the granules.

The examination of a transparent slice gives a good notion of the manner in which the components of the chalk are arranged, and of their relative proportions. But, by rubbing up some chalk with a brush in water and then pouring off the milky fluid, so as to obtain sediments of different degrees of fineness, the granules and the minute rounded bodies may be pretty well separated from one another, and submitted to microscopic examination, either as opaque or as transparent objects. By combining the views obtained in these various methods, each of the rounded bodies may be proved to be a beautifully-constructed calcareous fabric, made up of a number of chambers, communicating freely with one another. The chambered bodies are of various forms. One of the commonest is something like a badly-grown raspberry, being formed of a number of nearly globular chambers of different sizes congregated together. It is called *Globigerina*, and some specimens of chalk consist of little else than *Globigerinæ* and granules. Let us fix our attention upon the *Globigerina*. It is the spoor of the game we are tracking. If we can learn what it is and what are the conditions of its existence, we shall see our way to the origin and past history of the chalk.

A suggestion which may naturally enough present itself is, that these curious bodies are the result of some process of aggregation which has taken place in the carbonate of lime; that, just as in winter, the rime on our windows simulates the most delicate and elegantly arborescent foliage —proving that the mere mineral water may, under certain conditions, assume the outward form of organic bodies—so this mineral substance, carbonate of lime, hidden away in the bowels of the earth, has taken the shape of these chambered bodies. I am not raising a merely fanciful and unreal objection. Very learned men, in former days, have even entertained the notion that all the formed things found in rocks are of this nature; and if no such conception is at present held to be admissible, it is because long and varied experience has now shown that mineral matter never does assume the form and structure we find in fossils. If any one were to try to persuade you that an oyster-shell (which is also chiefly composed of carbonate of lime) had crystallized out of sea-water, I suppose you would laugh at the absurdity. Your laughter would be justified by the fact that all experience tends to show that oyster-shells are formed by the agency of oysters, and in no other way. And if there were no better reasons, we should be justified, on like grounds, in believing that *Globigerina* is not the product of anything but vital activity.

Happily, however, better evidence in proof of the organic nature of the *Globigerinæ* than that of analogy is forthcoming. It so happens that calcareous skeletons, exactly similar to the *Globigerinæ* of the chalk, are being formed, at the present moment, by minute living creatures, which flourish in multitudes, literally more numerous than the sands of the sea-shore, over a large extent of that part of the earth's surface which is covered by the ocean.

The history of the discovery of these living *Globigerinæ*, and of the part which they play in rock building, is singular enough. It is a discovery which, like others of no less scientific importance, has arisen, incidentally, out of work devoted to very different and exceedingly practical interests. When men first took to the sea, they speedily learned to look out for shoals and rocks; and the more the burthen of their ships increased, the more imperatively necessary it became for sailors to ascertain with precision the depth of the waters they traversed. Out of this necessity grew the use of the lead and sounding line; and, ultimately, marine-surveying, which is the recording of the form of coasts and of the depth of the sea, as ascertained by the sounding-lead, upon charts.

At the same time, it became desirable to ascertain and to indicate the nature of the sea-bottom, since this circumstance greatly affects its goodness

as holding ground for anchors. Some ingenious
tar, whose name deserves a better fate than the
oblivion into which it has fallen, attained this
object by "arming" the bottom of the lead with
a lump of grease, to which more or less of the
sand or mud, or broken shells, as the case might
be, adhered, and was brought to the surface. But,
however well adapted such an apparatus might
be for rough nautical purposes, scientific accuracy
could not be expected from the armed lead, and
to remedy its defects (especially when applied to
sounding in great depths) Lieut. Brooke, of the
American Navy, some years ago invented a most
ingenious machine, by which a considerable portion of the superficial layer of the sea-bottom can
be scooped out and brought up from any depth to
which the lead descends. In 1853, Lieut. Brooke
obtained mud from the bottom of the North
Atlantic, between Newfoundland and the Azores,
at a depth of more than 10,000 feet, or two miles,
by the help of this sounding apparatus. The
specimens were sent for examination to Ehrenberg
of Berlin, and to Bailey of West Point, and those
able microscopists found that this deep-sea mud
was almost entirely composed of the skeletons of
living organisms—the greater proportion of these
being just like the *Globigerinæ* already known to
occur in the chalk.

Thus far, the work had been carried on simply
in the interests of science, but Lieut. Brooke's

method of sounding acquired a high commercial value, when the enterprise of laying down the telegraph-cable between this country and the United States was undertaken. For it became a matter of immense importance to know, not only the depth of the sea over the whole line along which the cable was to be laid, but the exact nature of the bottom, so as to guard against chances of cutting or fraying the strands of that costly rope. The Admiralty consequently ordered Captain Dayman, an old friend and shipmate of mine, to ascertain the depth over the whole line of the cable, and to bring back specimens of the bottom. In former days, such a command as this might have sounded very much like one of the impossible things which the young Prince in the Fairy Tales is ordered to do before he can obtain the hand of the Princess. However, in the months of June and July, 1857, my friend performed the task assigned to him with great expedition and precision, without, so far as I know, having met with any reward of that kind. The specimens or Atlantic mud which he procured were sent to me to be examined and reported upon.[1]

---

[1] See Appendix to Captain Dayman's *Deep-sea Soundings in the North Atlantic Ocean between Ireland and Newfoundland, made in H. M. S. "Cyclops."* Published by order of the Lords Commissioners of the Admiralty, 1858. They have since formed the subject of an elaborate Memoir by Messrs. Parker and Jones, published in the *Philosophical Transactions* for 1865.

The result of all these operations is, that we know the contours and the nature of the surface-soil covered by the North Atlantic for a distance of 1,700 miles from east to west, as well as we know that of any part of the dry land. It is a prodigious plain—one of the widest and most even plains in the world. If the sea were drained off, you might drive a waggon all the way from Valentia, on the west coast of Ireland, to Trinity Bay, in Newfoundland. And, except upon one sharp incline about 200 miles from Valentia, I am not quite sure that it would even be necessary to put the skid on, so gentle are the ascents and descents upon that long route. From Valentia the road would lie down-hill for about 200 miles to the point at which the bottom is now covered by 1,700 fathoms of sea-water. Then would come the central plain, more than a thousand miles wide, the inequalities of the surface of which would be hardly perceptible, though the depth of water upon it now varies from 10,000 to 15,000 feet; and there are places in which Mont Blanc might be sunk without showing its peak above water. Beyond this, the ascent on the American side commences, and gradually leads, for about 300 miles, to the Newfoundland shore.

Almost the whole of the bottom of this central plain (which extends for many hundred miles in a north and. south direction) is covered by a fine mud, which, when brought to the surface, dries into a greyish white friable substance. You can write with this on a blackboard, if you are so inclined ; and, to the eye, it is quite like very soft, grayish chalk. Examined chemically, it proves to be composed almost wholly of carbonate of lime ; and if you make a section of it, in the same way as that of the piece of chalk was made, and view it with the microscope, it presents innumerable *Globigerinæ* embedded in a granular matrix. Thus this deep-sea mud is substantially chalk. I say substantially, because there are a good many minor differences ; but as these have no bearing on the question immediately before us,—which is the nature of the *Globigerinæ* of the chalk,—it is unnecessary to speak of them.

Globigerinæ of every size, from the smallest to the largest, are associated together in the Atlantic mud, and the chambers of many are filled by a soft animal matter. This soft substance is, in fact, the remains of the creature to which the *Globigerina* shell, or rather skeleton, owes its existence—and which is an animal of the simplest imaginable description. It is, in fact, a mere particle of living jelly, without defined parts of any kind—without a mouth, nerves, muscles, or distinct organs, and only manifesting its vitality to ordinary observation by thrusting out and retracting from all parts of its surface, long filamentous processes, which serve for arms and legs. Yet this amorphous particle, devoid of everything which, in the higher animals, we call organs, is capable of feeding, growing, and multiplying; of separating from the ocean the small proportion of carbonate of lime which is dissolved in sea-water; and of building up that substance into a skeleton for itself, according to a pattern which can be imitated by no other known agency.

The notion that animals can live and flourish in the sea, at the vast depths from which apparently living *Globigerinæ* have been brought up, does not agree very well with our usual conceptions respecting the conditions of animal life; and it is not so absolutely impossible as it might at first sight appear to be, that the *Globigerinæ* of the Atlantic sea-bottom do not live and die where they are found.

As I have mentioned, the soundings from the great Atlantic plain are almost entirely made up of *Globigerinæ*, with the granules which have been mentioned, and some few other calcareous shells; but a small percentage of the chalky mud—perhaps at most some five per cent. of it—is of a different nature, and consists of shells and skeletons composed of silex, or pure flint. These silicious bodies belong partly to the lowly vegetable organisms which are called *Diatomaceæ*, and partly to the minute, and extremely simple, animals, termed *Radiolaria*. It is quite certain that these creatures do not live at the bottom of the ocean, but at its surface—where they may be obtained in prodigious numbers by the use of a properly constructed net. Hence it follows that these silicious organisms, though they are not heavier than the lightest dust, must have fallen, in some cases, through fifteen thousand feet of water, before they reached their final resting-place on the ocean floor. And considering how large a surface these bodies expose in proportion to their weight, it is probable that they occupy a great length of time in making their burial journey from the surface of the Atlantic to the bottom.

But if the *Radiolaria* and Diatoms are thus rained upon the bottom of the sea, from the superficial layer of its waters in which they pass their lives, it is obviously possible that the *Globigerinæ* may be similarly derived; and if they were so, it would be much more easy to understand how they obtain their supply of food than it is at present. Nevertheless, the positive and negative evidence all points the other way. The skeletons of the full-grown, deep-sea *Globigerinæ* are so remarkably solid and heavy in proportion to their surface as to seem little fitted for floating; and, as a matter of fact, they are not to be found along with the Diatoms and *Radiolaria* in the uppermost stratum of the open ocean. It has been observed, again, that the abundance of *Globigerinæ*, in proportion to other organisms, of like kind, increases with the depth of the sea; and that deep-water *Globigerinæ* are larger than those which live in shallower parts of the sea; and such facts negative the supposition that these organisms have been swept by currents from the shallows into the deeps of the Atlantic. It therefore seems to be hardly doubtful that these wonderful creatures live and die at the depths in which they are found.[1]

However, the important points for us are, that the living *Globigerinæ* are exclusively marine animals, the skeletons of which abound at the bottom of deep seas; and that there is not a shadow of reason for believing that the habits of the *Globigerinæ* of the chalk differed from those of the existing species. But if this be true, there is no escaping the conclusion that the chalk itself is the dried mud of an ancient deep sea.

In working over the soundings collected by Captain Dayman, I was surprised to find that many of what I have called the "granules" of that mud were not, as one might have been tempted

---

[1] During the cruise of H.M.S. *Bulldog*, commanded by Sir Leopold M'Clintock, in 1860, living star-fish were brought up, clinging to the lowest part of the sounding-line, from a depth of 1,260 fathoms, midway between Cape Farewell, in Greenland, and the Rockall banks. Dr. Wallich ascertained that the sea-bottom at this point consisted of the ordinary *Globigerina* ooze, and that the stomachs of the star-fishes were full of *Globigerinæ*. This discovery removes all objections to the existence of living *Globigerinæ* at great depths, which are based upon the supposed difficulty of maintaining animal life under such conditions; and it throws the burden of proof upon those who object to the supposition that the *Globigerinæ* live and die where they are found.

to think at first, the mere powder and waste of *Globigerinæ*, but that they had a definite form and size. I termed these bodies "*coccoliths*," and doubted their organic nature. Dr. Wallich verified my observation, and added the interesting dis-covery that, not unfrequently, bodies similar to these "coccoliths" were aggregated together into spheroids, which he termed "*coccospheres.*" So far as we knew, these bodies, the nature of which is extremely puzzling and problematical, were peculiar to the Atlantic soundings. But, a few years ago, Mr. Sorby, in making a careful examina-tion of the chalk by means of thin sections and otherwise, observed, as Ehrenberg had done before him, that much of its granular basis possesses a definite form. Comparing these formed particles with those in the Atlantic soundings, he found the two to be identical; and thus proved that the chalk, like the surroundings, contains these mys-terious coccoliths and coccospheres. Here was a further and most interesting confirmation, from internal evidence, of the essential identity of the chalk with modern deep-sea mud. *Globigerinæ*, coccoliths, and coccospheres are found as the chief constituents of both, and testify to the general similarity of the conditions under which both have been formed.[1]

The evidence furnished by the hewing, facing,

[1] I have recently traced out the development of the "cocco-liths" from a diameter of $\frac{1}{7000}$th of an inch up to their largest

and superposition of the stones of the Pyramids, that these structures were built by men, has no greater weight than the evidence that the chalk was built by *Globigerinæ;* and the belief that those ancient pyramid-builders were terrestrial and air-breathing creatures like ourselves, is not better based than the conviction that the chalk-makers lived in the sea.   But as our belief in the building of the Pyramids by men is not only grounded on the internal evidence afforded by these structures, but gathers strength from multitudinous collateral proofs, and is clinched by the total absence of any reason for a contrary belief; so the evidence drawn from the *Globigerinæ* that the chalk is an ancient sea-bottom, is fortified by innumerable independent lines of evidence; and our belief in the truth of the conclusion to which all positive testimony tends, receives the like negative justification from the fact that no other hypothesis has a shadow of foundation.

It may be worth while briefly to consider a few of these collateral proofs that the chalk was deposited at the bottom of the sea.   The great mass of the chalk is composed, as we have seen, of the skeletons of *Globigerinæ,* and other simple organisms, imbedded in granular matter.   Here and there, however, this hardened mud of the

size (which is about $\frac{1}{1000}$th), and no longer doubt that they are produced by independent organisms, which, like the *Globigerinæ*, live and die at the bottom of the sea.

ancient sea reveals the remains of higher animals which have lived and died, and left their hard parts in the mud, just as the oysters die and leave their shells behind them, in the mud of the present seas.

There are, at the present day, certain groups of animals which are never found in fresh waters, being unable to live anywhere but in the sea. Such are the corals; those corallines which are called *Polyzoa;* those creatures which fabricate the lamp-shells, and are called *Brachiopoda;* the pearly *Nautilus,* and all animals allied to it; and all the forms of sea-urchins and star-fishes. Not only are all these creatures confined to salt water at the present day; but, so far as our records of the past go, the conditions of their existence have been the same: hence, their occurrence in any deposit is as strong evidence as can be obtained, that that deposit was formed in the sea. Now the remains of animals of all the kinds which have been enumerated, occur in the chalk, in greater or less abundance; while not one of those forms of shell-fish which are characteristic of fresh water has yet been observed in it.

When we consider that the remains of more than three thousand distinct species of aquatic animals have been discovered among the fossils of the chalk, that the great majority of them are of such forms as are now met with only in the sea, and that there is no reason to believe that any

one of them inhabited fresh water—the collateral
evidence that the chalk represents an ancient sea-
bottom acquires as great force as the proof
derived from the nature of the chalk itself. I
think you will now allow that I did not overstate
my case when I asserted that we have as strong
grounds for believing that all the vast area of
dry land, at present occupied by the chalk, was
once at the bottom of the sea, as we have for any
matter of history whatever; while there is no
justification for any other belief.

No less certain it is that the time during which
the countries we now call south-east England,
France, Germany, Poland, Russia, Egypt, Arabia,
Syria, were more or less completely covered by a
deep sea, was of considerable duration. We have
already seen that the chalk is, in places, more
than a thousand feet thick. I think you will
agree with me, that it must have taken some
time for the skeletons of animalcules of a
hundredth of an inch in diameter to heap up
such a mass as that. I have said that throughout the thickness of the chalk the remains of
other animals are scattered. These remains are
often in the most exquisite state of preservation.
The valves of the shell-fishes are commonly
adherent; the long spines of some of the sea-
urchins, which would be detached by the smallest
jar, often remain in their places. In a word, it is
certain that these animals have lived and died

when the place which they now occupy was the surface of as much of the chalk as had then been deposited; and that each has been covered up by the layer of *Globigerina* mud, upon which the creatures imbedded a little higher up have, in like manner, lived and died. But some of these remains prove the existence of reptiles of vast size in the chalk sea. These lived their time, and had their ancestors and descendants, which assuredly implies time, reptiles being of slow growth.

There is more curious evidence, again, that the process of covering up, or, in other words, the deposit of *Globigerina* skeletons, did not go on very fast. It is demonstrable that an animal of the cretaceous sea might die, that its skeleton might lie uncovered upon the sea-bottom long enough to lose all its outward coverings and appendages by putrefaction; and that, after this had happened, another animal might attach itself to the dead and naked skeleton, might grow to maturity, and might itself die before the calcareous mud had buried the whole.

Cases of this kind are admirably described by Sir Charles Lyell. He speaks of the frequency with which geologists find in the chalk a fossilized sea-urchin, to which is attached the lower valve of a *Crania*. This is a kind of shell-fish, with a shell composed of two pieces, of which, as in the oyster, one is fixed and the other free.

"The upper valve is almost invariably wanting, though occasionally found in a perfect state of preservation in the white chalk at some distance. In this case, we see clearly that the sea-urchin first lived from youth to age, then died and lost its spines, which were carried away. Then the young *Crania* adhered to the bared shell, grew and perished in its turn; after which, the upper valve was separated from the lower, before the Echinus became enveloped in chalky mud."[1]

A specimen in the Museum of Practical Geology, in London, still further prolongs the period which must have elapsed between the death of the sea-urchin, and its burial by the *Globigerinæ*. For the outward face of the valve of a *Crania*, which is attached to a sea-urchin, (*Micraster*), is itself overrun by an incrusting coralline, which spreads thence over more or less of the surface of the sea-urchin. It follows that, after the upper valve of the *Crania* fell off, the surface of the attached valve must have remained exposed long enough to allow of the growth of the whole coralline, since corallines do not live embedded in mud.[1]

The progress of knowledge may, one day, enable us to deduce from such facts as these the maximum rate at which the chalk can have accumulated, and thus to arrive at the minimum

---

[1] *Elements of Geology*, by Sir Charles Lyell, Bart. F.R.S., p. 23.

duration of the chalk period. Suppose that the valve of the *Crania* upon which a coralline has fixed itself in the way just described, is so attached to the sea-urchin that no part of it is more than an inch above the face upon which the sea-urchin rests. Then, as the coralline could not have fixed itself, if the *Crania* had been covered up with chalk mud, and could not have lived had itself been so covered, it follows, that an inch of chalk mud could not have accumulated within the time between the death and decay of the soft parts of the sea-urchin and the growth of the coralline to the full size which it has attained. If the decay of the soft parts of the sea-urchin; the attachment, growth to maturity, and decay of the *Crania*; and the subsequent attachment and growth of the coralline, took a year (which is a low estimate enough), the accumulation of the inch of chalk must have taken more than a year: and the deposit of a thousand feet of chalk must, consequently, have taken more than twelve thousand years.

The foundation of all this calculation is, of course, a knowledge of the length of time the *Crania* and the coralline needed to attain their full size; and, on this head, precise knowledge is at present wanting. But there are circumstances which tend to show, that nothing like an inch of chalk has accumulated during the life of a *Crania*; and, on any probable estimate of the length of

that life, the chalk period must have had a much
longer duration than that thus roughly assigned
to it.

Thus, not only is it certain that the chalk
is the mud of an ancient sea-bottom; but it is no
less certain, that the chalk sea existed during an
extremely long period, though we may not be
prepared to give a precise estimate of the length
of that period in years. The relative duration is
clear, though the absolute duration may not be
definable. The attempt to affix any precise date
to the period at which the chalk sea began, or
ended, its existence, is baffled by difficulties of the
same kind. But the relative age of the cretaceous
epoch may be determined with as great ease
and certainty as the long duration of that epoch.
You will have heard of the interesting discoveries recently made, in various parts of Western
Europe, of flint implements, obviously worked into
shape by human hands, under circumstances which
show conclusively that man is a very ancient
denizen of these regions. It has been proved that
the whole populations of Europe, whose existence
has been revealed to us in this way, consisted of
savages, such as the Esquimaux are now; that, in
the country which is now France, they hunted the
reindeer, and were familiar with the ways of the
mammoth and the bison. The physical geography
of France was in those days different from what it

is now—the river Somme, for instance, having cut its bed a hundred feet deeper between that time and this ; and, it is probable, that the climate was more like that of Canada or Siberia, than that of Western Europe.

The existence of these people is forgotten even in the traditions of the oldest historical nations. The name and fame of them had utterly vanished until a few years back ; and the amount of physical change which has been effected since their day renders it more than probable that, venerable as are some of the historical nations, the workers of the chipped flints of Hoxne or of Amiens are to them, as they are to us, in point of antiquity. But, if we assign to these hoar relics of long-vanished generations of men the greatest age that can possibly be claimed for them, they are not older than the drift, or boulder clay, which, in comparison with the chalk, is but a very juvenile deposit. You need go no further than your own sea-board for evidence of this fact. At one of the most charming spots on the coast of Norfolk, Cromer, you will see the boulder clay forming a vast mass, which lies upon the chalk, and must consequently have come into existence after it. Huge boulders of chalk are, in fact, included in the clay, and have evidently been brought to the position they now occupy by the same agency as that which has planted blocks of syenite from Norway side by side with them.

The chalk, then, is certainly older than the
boulder clay. If you ask how much, I will again
take you no further than the same spot upon your
own coasts for evidence. I have spoken of the
boulder clay and drift as resting upon the chalk.
That is not strictly true. Interposed between the
chalk and the drift is a comparatively insignificant layer, containing vegetable matter. But that
layer tells a wonderful history. It is full of stumps
of trees standing as they grew. Fir-trees are there
with their cones, and hazel-bushes with their
nuts; there stand the stools of oak and yew trees,
beeches and alders. Hence this stratum is appro-
priately called the "forest-bed."

It is obvious that the chalk must have been
upheaved and converted into dry land, before the
timber trees could grow upon it. As the bolls of
some of these trees are from two to three feet in
diameter, it is no less clear that the dry land thus
formed remained in the same condition for long
ages. And not only do the remains of stately
oaks and well-grown firs testify to the duration of
this condition of things, but additional evidence to
the same effect is afforded by the abundant remains of elephants, rhinoceroses, hippopotamuses,
and other great wild beasts, which it has yielded
to the zealous search of such men as the Rev. Mr.
Gunn. When you look at such a collection as he
has formed, and bethink you that these elephan-
tine bones did veritably carry their owners about,

and these great grinders crunch, in the dark woods
of which the forest-bed is now the only trace, it is
impossible not to feel that they are as good
evidence of the lapse of time as the annual rings
of the tree stumps.

Thus there is a writing upon the wall of cliffs
at Cromer, and whoso runs may read it. It tells
us, with an authority which cannot be impeached,
that the ancient sea-bed of the chalk sea was
raised up, and remained dry land, until it was
covered with forest, stocked with the great game the
spoils of which have rejoiced your geologists. How
long it remained in that condition cannot be said ;
but " the whirligig of time brought its revenges "
in those days as in these. That dry land, with
the bones and teeth of generations of long-lived
elephants, hidden away among the gnarled roots
and dry leaves of its ancient trees, sank gradually
to the bottom of the icy sea, which covered it with
huge masses of drift and boulder clay. Sea-beasts,
such as the walrus, now restricted to the extreme
north, paddled about where birds had twittered
among the topmost twigs of the fir-trees. How
long this state of things endured we know not,
but at length it came to an end. The upheaved
glacial mud hardened into the soil of modern
Norfolk. Forests grew once more, the wolf and
the beaver replaced the reindeer and the elephant ;
and at length what we call the history of England
dawned.

Thus you have, within the limits of your own

county, proof that the chalk can justly claim a very much greater antiquity than even the oldest physical traces of mankind. But we may go further and demonstrate, by evidence of the same authority as that which testifies to the existence of the father of men, that the chalk is vastly older than Adam himself. The Book of Genesis informs us that Adam, immediately upon his creation, and before the appearance of Eve, was placed in the Garden of Eden. The problem of the geographical position of Eden has greatly vexed the spirits of the learned in such matters, but there is one point respecting which, so far as I know, no commentator has ever raised a doubt. This is, that of the four rivers which are said to run out of it, Euphrates and Hiddekel are identical with the rivers now known by the names of Euphrates and Tigris. But the whole country in which these mighty rivers take their origin, and through which they run, is composed of rocks which are either of the same age as the chalk, or of later date. So that the chalk must not only have been formed, but, after its formation, the time required for the deposit of these later rocks, and for their upheaval into dry land, must have elapsed, before the smallest brook which feeds the swift stream of "the great river, the river of Babylon," began to flow.

Thus, evidence which cannot be rebutted, and which need not be strengthened, though if time permitted I might indefinitely increase its quantity, compels you to believe that the earth, from the time of the chalk to the present day, has been the theatre of a series of changes as vast in their amount, as they were slow in their progress. The area on which we stand has been first sea and then land, for at least four alternations; and has remained in each of these conditions for a period of great length.

Nor have these wonderful metamorphoses of sea into land, and of land into sea, been confined to one corner of England. During the chalk period, or " cretaceous epoch," not one of the present great physical features of the globe was in existence. Our great mountain ranges, Pyrenees, Alps, Himalayas, Andes, have all been upheaved since the chalk was deposited, and the cretaceous sea flowed over the sites of Sinai and Ararat. All this is certain, because rocks of cretaceous, or still later, date have shared in the elevatory movements which gave rise to these mountain chains; and may be found perched up, in some cases, many thousand feet high upon their flanks. And evidence of equal cogency demonstrates that, though, in Norfolk, the forest-bed rests directly upon the chalk, yet it does so, not because the period at which the forest grew immediately followed that at which the chalk was formed, but because an immense lapse of time, represented elsewhere by thousands of feet of rock, is not indicated at Cromer.

I must ask you to believe that there is no less conclusive proof that a still more prolonged succession of similar changes occurred, before the chalk was deposited. Nor have we any reason to think that the first term in the series of these changes is known. The oldest sea-beds preserved to us are sands, and mud, and pebbles, the wear and tear of rocks which were formed in still older oceans.

But, great as is the magnitude of these physical changes of the world, they have been accompanied by a no less striking series of modifications in its living inhabitants. All the great classes of animals, beasts of the field, fowls of the air, creeping things, and things which dwell in the waters, flourished upon the globe long ages before the chalk was deposited. Very few, however, if any, of these ancient forms of animal life were identical with those which now live. Certainly not one of the higher animals was of the same species as any of those now in existence. The beasts of the field, in the days before the chalk, were not our beasts of the field, nor the fowls of the air such as those which the eye of men has seen flying, unless his antiquity dates infinitely further back than we at present surmise. If we could be carried back into those times, we should be as one suddenly set down in Australia before it was colonized. We should see mammals, birds, reptiles, fishes, insects, snails, and the like, clearly recognizable as such, and yet not one of them would be just the same as those with which we are familiar, and many would be extremely different.

From that time to the present, the population of the world has undergone slow and gradual, but incessant, changes. There has been no grand catastrophe—no destroyer has swept away the forms of life of one period, and replaced them by a totally new creation : but one species has vanished and another has taken its place ; creatures of one type of structure have diminished, those of another have increased, as time has passed on. And thus, while the differences between the living creatures of the time before the chalk and those of the present day appear startling, if placed side by side, we are led from one to the other by the most gradual progress, if we follow the course of Nature through the whole series of those relics of her operations which she has left behind. It is by the population of the chalk sea that the ancient and the modern inhabitants of the world are most completely connected. The groups which are dying out flourish, side by side, with the groups which are now the dominant forms of life. Thus the chalk contains remains of those strange flying and swimming reptiles, the pterodactyl, the ichthyosaurus, and the plesiosaurus, which are found in no later deposits, but abounded in preceding ages. The chambered shells called ammonites and belemnites, which are so characteristic of the period preceding the cretaceous, in like manner die with it.

But, amongst these fading remainders of a previous state of things, are some very modern forms of life, looking like Yankee pedlars among a tribe of Red Indians. Crocodiles of modern type appear; bony fishes, many of them very similar to existing species, almost supplant the forms of fish which predominate in more ancient seas; and many kinds of living shell-fish first become known to us in the chalk. The vegetation acquires a modern aspect. A few living animals are not even distinguishable as species, from those which existed at that remote epoch. The *Globigerina* of the present day, for example, is not different specifically from that of the chalk; and the same may be said of many other *Foraminifera*. I think it probable that critical and unprejudiced examination will show that more than one species of much higher animals have had a similar longevity; but the only example which I can at present give confidently is the snake's-head lampshell (*Terebratulina caput serpentis*), which lives in our English seas and abounded (as *Terebratulina striata* of authors) in the chalk.

The longest line of human ancestry must hide its diminished head before the pedigree of this insignificant shell-fish. We Englishmen are proud to have an ancestor who was present at the

Battle of Hastings. The ancestors of *Terebratulina caput serpentis* may have been present at a battle of *Ichthyosauria* in that part of the sea which, when the chalk was forming, flowed over the site of Hastings. While all around has changed, this *Terebratulina* has peacefully propagated its species from generation to generation, and stands to this day, as a living testimony to the continuity of the present with the past history of the globe.

Up to this moment I have stated, so far as I know, nothing but well-authenticated facts, and the immediate conclusions which they force upon the mind. But the mind is so constituted that it does not willingly rest in facts and immediate causes, but seeks always after a knowledge of the remoter links in the chain of causation.

Taking the many changes of any given spot of the earth's surface, from sea to land and from land to sea, as an established fact, we cannot refrain from asking ourselves how these changes have occurred. And when we have explained them— as they must be explained—by the alternate slow movements of elevation and depression which have affected the crust of the earth, we go still further back, and ask, Why these movements?

I am not certain that any one can give you a satisfactory answer to that question. Assuredly I cannot. All that can be said, for certain, is, that such movements are part of the ordinary course of nature, inasmuch as they are going on at the present time. Direct proof may be given, that some parts of the land of the northern hemisphere are at this moment insensibly rising and others insensibly sinking; and there is indirect, but perfectly satisfactory, proof, that an enormous area now covered by the Pacific has been deepened thousands of feet, since the present inhabitants of that sea came into existence. Thus there is not a shadow of a reason for believing that the physical changes of the globe, in past times, have been effected by other than natural causes. Is there any more reason for believing that the concomitant modifications in the forms of the living inhabitants of the globe have been brought about in other ways?

Before attempting to answer this question, let us try to form a distinct mental picture of what has happened in some special case. The crocodiles are animals which, as a group, have a very vast antiquity. They abounded ages before the chalk was deposited; they throng the rivers in warm climates, at the present day. There is a difference in the form of the joints of the back-bone, and in some minor particulars, between the crocodiles of the present epoch and those which lived before the chalk; but, in the cretaceous epoch, as I have already mentioned, the crocodiles had assumed the modern type of structure. Notwithstanding this, the crocodiles of the chalk are not identically the same as those which lived in the times called " older tertiary," which succeeded the cretaceous epoch; and the crocodiles of the older tertiaries are not identical with those of the newer tertiaries, nor are these identical with existing forms. I leave open the question whether particular species may have lived on from epoch to epoch. But each epoch has had its peculiar crocodiles; though all, since the chalk, have belonged to the modern type, and differ simply in their proportions, and in such structural particulars as are discernible only to trained eyes.

How is the existence of this long succession of different species of crocodiles to be accounted for? Only two suppositions seem to be open to us— Either each species of crocodile has been specially created, or it has arisen out of some pre-existing form by the operation of natural causes. Choose your hypothesis; I have chosen mine. I can find no warranty for believing in the distinct creation of a score of successive species of crocodiles in the course of countless ages of time. Science gives no countenance to such a wild fancy; nor can even the perverse ingenuity of a commentator pretend to discover this sense, in the simple words in which the writer of Genesis records the pro- ceedings of the fifth and six days of the Creation.

On the other hand, I see no good reason for doubting the necessary alternative, that all these varied species have been evolved from pre-existing crocodilian forms, by the operation of causes as completely a part of the common order of nature as those which have effected the changes of the inorganic world. Few will venture to affirm that the reasoning which applies to crocodiles loses its force among other animals, or among plants. If one series of species has come into existence by the operation of natural causes, it seems folly to deny that all may have arisen in the same way.

A small beginning has led us to a great ending. If I were to put the bit of chalk with which we started into the hot but obscure flame of burning hydrogen, it would presently shine like the sun. It seems to me that this physical metamorphosis is no false image of what has been the result of our subjecting it to a jet of fervent, though no-wise brilliant, thought to-night. It has become luminous, and its clear rays, penetrating the abyss of the remote past, have brought within our ken some stages of the evolution of the earth. And in the shifting " without haste, but without rest " of the land and sea, as in the endless variation of the forms assumed by living beings, we have observed nothing but the natural product of the forces originally possessed by the substance of the universe.

II

THE PROBLEMS OF THE DEEP SEA

[1873]

ON the 21st of December, 1872, H.M.S. *Challenger*, an eighteen gun corvette, of 2,000 tons burden, sailed from Portsmouth harbour for a three, or perhaps four, years' cruise. No man-of-war ever left that famous port before with so singular an equipment. Two of the eighteen sixty-eight pounders of the *Challenger's* armament remained to enable her to speak with effect to sea-rovers, haply devoid of any respect for science, in the remote seas for which she is bound ; but the main-deck was, for the most part, stripped of its war-like gear, and fitted up with physical, chemical, and biological laboratories; photography had its dark cabin ; while apparatus for dredging, trawling, and sounding ; for photo-meters and for thermometers, filled the space formerly occupied by guns and gun-tackle, pistols and cutlasses.

The crew of the *Challenger* match her fittings. Captain Nares, his officers and men, are ready to look after the interests of hydrography, work the ship, and, if need be, fight her as seamen should; while there is a staff of scientific civilians, under the general direction of Dr. Wyville Thomson, F.R.S. (Professor of Natural History in Edinburgh University by rights, but at present detached for duty *in partibus*), whose business it is to turn all the wonderfully packed stores of appliances to account, and to accumulate, before the ship returns to England, such additions to natural knowledge as shall justify the labour and cost involved in the fitting out and maintenance of the expedition.

Under the able and zealous superintendence of the Hydrographer, Admiral Richards, every precaution which experience and forethought could devise has been taken to provide the expedition with the material conditions of success; and it would seem as if nothing short of wreck or pestilence, both most improbable contingencies, could prevent the *Challenger* from doing splendid work, and opening up a new era in the history of scientific voyages.

The dispatch of this expedition is the culmination of a series of such enterprises, gradually increasing in magnitude and importance, which the Admiralty, greatly to its credit, has carried out for some years past; and the history of which is given by Dr. Wyville Thomson in the beautifully illus-

trated volume entitled "The Depths of the Sea,"
published since his departure.

"In the spring of the year 1868, my friend Dr. W. B. Carpenter, at that time one of the Vice-Presidents of the Royal
Society, was with me in Ireland, where we were working out
together the structure and development of the Crinoids. I had
long previously had a profound conviction that the land of
promise for the naturalist, the only remaining region where
there were endless novelties of extraordinary interest ready to
the hand which had the means of gathering them, was the
bottom of the deep sea. I had even had a glimpse of some of
these treasures, for I had seen, the year before, with Prof. Sars,
the forms which I have already mentioned dredged by his son
at a depth of 300 to 400 fathoms off the Loffoten Islands. I
propounded my views to my fellow-labourer, and we discussed
the subject many times over our microscopes. I strongly urged
Dr. Carpenter to use his influence at head-quarters to induce the
Admiralty, probably through the Council of the Royal Society,
to give us the use of a vessel properly fitted with dredging gear
and all necessary scientific apparatus, that many heavy questions
as to the state of things in the depths of the ocean, which were
still in a state of uncertainty, might be definitely settled.
After full consideration, Dr. Carpenter promised his hearty co-
operation, and we agreed that I should write to him on his
return to London, indicating generally the results which I anticipated, and sketching out what I conceived to be a promising
line of inquiry. The Council of the Royal Society warmly
supported the proposal ; and I give here in chronological order
the short and eminently satisfactory correspondence which led
to the Admiralty placing at the disposal of Dr. Carpenter and
myself the gunboat *Lightning*, under the command of Staff-
Commander May, R.N., in the summer of 1868, for a trial
cruise to the North of Scotland, and afterwards to the much
wider surveys in H.M.S. *Porcupine*, Captain Calver, R.N.,
which were made with the additional association of Mr. Gwyn
Jeffreys, in the summers of the years 1869 and 1870."[1]

---

[1] *The Depths of the Sea*, pp. 49-50.

Plain men may be puzzled to understand why Dr. Wyville Thomson, not being a cynic, should relegate the " Land of Promise " to the bottom of the deep sea; they may still more wonder what manner of " milk and honey " the *Challenger* expects to find; and their perplexity may well rise to its maximum, when they seek to divine the manner in which that milk and honey are to be got out of so inaccessible a Canaan. I will, therefore, endeavour to give some answer to these questions in an order the reverse of that in which I have stated them.

Apart from hooks, and lines, and ordinary nets, fishermen have, from time immemorial, made use of two kinds of implements for getting at sea-creatures which live beyond tide-marks—these are the " dredge " and the " trawl." The dredge is used by oyster-fishermen. Imagine a large bag, the mouth of which has the shape of an elongated parallelogram, and is fastened to an iron frame of the same shape, the two long sides of this rim being fashioned into scrapers. Chains attach the ends of the frame to a stout rope, so that when the bag is dragged along by the rope the edge of one of the scrapers rests on the ground, and scrapes whatever it touches into the bag. The oyster-dredger takes one of these machines in his boat, and when he has reached the oyster-bed the dredge is tossed overboard; as soon as it has sunk to the bottom the rope is paid out sufficiently to prevent it from pulling the dredge directly upwards, and is then made fast while the boat goes ahead. The dredge is thus dragged along and scrapes oysters and other sea-animals and plants, stones, and mud into the bag. When the dredger judges it to be full he hauls it up, picks out the oysters, throws the rest overboard, and begins again.

Dredging in shallow water, say ten to twenty fathoms, is an easy operation enough ; but the deeper the dredger goes, the heavier must be his vessel, and the stouter his tackle, while the operation of hauling up becomes more and more laborious. Dredging in 150 fathoms is very hard work, if it has to be carried on by manual labour ; but by the use of the donkey-engine to supply power,[1] and of the contrivances known as " accumulators," to diminish the risk of snapping the dredge rope by the rolling and pitching of the vessel, the dredge has been worked deeper and deeper, until at last, on the 22nd of July, 1869, H.M.S. *Porcupine* being in the Bay of Biscay, Captain Calver, her commander, performed the unprecedented feat of dredging in 2,435 fathoms, or 14,610 feet, a depth

---

[1] The emotional side of the scientific nature has its singularities. Many persons will call to mind a certain philosopher's tenderness over his watch—" the little creature "—which was so singularly lost and found again. But Dr. Wyville Thomson surpasses the owner of the watch in his loving-kindness towards a donkey-engine. " This little engine was the comfort of our lives. Once or twice it was overstrained, and then we pitied the willing little thing, panting like an overtaxed horse."

nearly equal to the height of Mont Blanc. The
dredge " was rapidly hauled on deck at one o'clock
in the morning of the 23rd, after an absence of
7¼ hours, and a journey of upwards of eight statute
miles," with a hundred weight and a half of solid
contents.

The trawl is a sort of net for catching those fish
which habitually live at the bottom of the sea,
such as soles, plaice, turbot, and gurnett. The
mouth of the net may be thirty or forty feet wide,
and one edge of its mouth is fastened to a beam
of wood of the same length. The two ends of the
beam are supported by curved pieces of iron,
which raise the beam and the edge of the net
which is fastened to it, for a short distance, while
the other edge of the mouth of the net trails upon
the ground. The closed end of the net has the
form of a great pouch; and, as the beam is
dragged along, the fish, roused from the bottom
by the sweeping of the net, readily pass into its
mouth and accumulate in the pouch at its end.
After drifting with the tide for six or seven hours
the trawl is hauled up, the marketable fish are
picked out, the others thrown away, and the trawl
sent overboard for another operation.

More than a thousand sail of well-found trawlers
are constantly engaged in sweeping the seas
around our coast in this way, and it is to them
that we owe a very large proportion of our supply
of fish. The difficulty of trawling, like that of

dredging, rapidly increases with the depth at which the operation is performed ; and, until the other day, it is probable that trawling at so great a depth as 100 fathoms was something unheard of. But the first news from the *Challenger* opens up new possibilities for the trawl.

Dr. Wyville Thomson writes (" Nature," March 20, 1873) :—

"For the first two or three hauls in very deep water off the coast of Portugal, the dredge came up filled with the usual 'Atlantic ooze,' tenacious and uniform throughout, and the work of hours, in sifting, gave the very smallest possible result. We were extremely anxious to get some idea of the general character of the Fauna, and particularly of the distribution of the higher groups ; and after various suggestions for modification of the dredge, it was proposed to try the ordinary trawl. We had a compact trawl, with a 15-feet beam, on board, and we sent it down off Cape St. Vincent at a depth of 600 fathoms. The experiment looked hazardous, but, to our great satisfaction, the trawl came up all right and contained, with many of the larger invertebrata, several fishes. . . . After the first attempt we tried the trawl several times at depths of 1090, 1525, and, finally, 2125 fathoms, and always with success."

To the coral-fishers of the Mediterranean, who seek the precious red coral, which grows firmly fixed to rocks at a depth of sixty to eighty fathoms, both the dredge and the trawl would be useless. They, therefore, have recourse to a sort of frame, to which are fastened long bundles of loosely netted hempen cord, and which is lowered by a rope to the depth at which the hempen cords can sweep over the surface of the rocks and break off the coral, which is brought up entangled in the cords. A similar contrivance has arisen out of the necessities of deep-sea exploration.

In the course of the dredging of the *Porcupine*, it was frequently found that, while few objects of interest were brought up within the dredge, many living creatures came up sticking to the outside of the dredge-bag, and even to the first few fathoms of the dredge-rope. The mouth of the dredge doubtless rapidly filled with mud, and thus the things it should have brought up were shut out. To remedy this inconvenience Captain Calver devised an arrangement not unlike that employed by the coral-fishers. He fastened half a dozen swabs, such as are used for drying decks, to the dredge. A swab is something like what a birch-broom would be if its twigs were made of long, coarse, hempen yarns. These dragged along after the dredge over the surface of the mud, and entangled the creatures living there—multitudes of which, twisted up in the strands of the swabs, were brought to the surface with the dredge. A further improvement was made by attaching a long iron bar to the bottom of the dredge bag, and fastening large bunches of teased-out hemp to the end of this bar. These "tangles" bring up immense quantities of such animals as have long arms, or spines, or prominences which readily become caught in the hemp, but they are very destructive to the fragile organisms which they

imprison; and, now that the trawl can be successfully worked at the greatest depths, it may be
expected to supersede them; at least, wherever
the ground is soft enough to permit of trawling.

It is obvious that between the dredge, the trawl,
and the tangles, there is little chance for any
organism, except such as are able to burrow
rapidly, to remain safely at the bottom of any part
of the sea which the *Challenger* undertakes to
explore.   And, for the first time in the history of
scientific exploration, we have a fair chance of learning what the population of the depths of the sea is
like in the most widely different parts of the world.

And now arises the next question.   The means
of exploration being fairly adequate, what forms
of life may be looked for at these vast depths?

The systematic study of the Distribution of
living beings is the most modern branch of Biological Science, and came into existence long after
Morphology and Physiology had attained a considerable development.   This naturally does not
imply that, from the time men began to observe
natural phenomena, they were ignorant of the fact
that the animals and plants of one part of the
world are different from those in other regions; or
that those of the hills are different from those of
the plains in the same region; or finally that
some marine creatures are found only in the
shallows, while others inhabit the deeps.   Nevertheless, it was only after the discovery of America

that the attention of naturalists was powerfully drawn to the wonderful differences between the animal population of the central and southern parts of the new world and that of those parts of the old world which lie under the same parallels of latitude. So far back as 1667 Abraham Mylius, in his treatise " De Animalium origine et migratione populorum," argues that, since there are innumerable species of animals in America which do not exist elsewhere, they must have been made and placed there by the Deity: Buffon no less forcibly insists upon the difference between the Faunæ of the old and new world. But the first attempt to gather facts of this order into a whole, and to coordinate them into a series of generalizations, or laws of Geographical Distribution, is not a century old, and is contained in the " Specimen Zoologiæ Geographicæ Quadrupedum Domicilia et Migrationes sistens," published, in 1777, by the learned Brunswick Professor, Eberhard Zimmermann, who illustrates his work by what he calls a " Tabula Zoographica," which is the oldest distributional map known to me.

In regard to matters of fact, Zimmermann's chief aim is to show that among terrestrial mammals, some occur all over the world, while others are restricted to particular areas of greater or smaller extent; and that the abundance of species follows temperature, being greatest in warm and least in cold climates. But marine animals,

he thinks, obey no such law.   The Arctic and
Atlantic seas, he says, are as full of fishes and
other animals as those of the tropics.   It is, therefore, clear that cold does not affect the dwellers
in the sea as it does land animals, and that this
must be the case follows from the fact that sea
water, "propter varias quas continet bituminis
spiritusque particulas," freezes with much more
difficulty than fresh water.   On the other hand,
the heat of the Equatorial sun penetrates but a
short distance below the surface of the ocean.
Moreover, according to Zimmermann, the incessant
disturbance of the mass of the sea by winds and
tides, so mixes up the warm and the cold that
life is evenly diffused and abundant throughout
the ocean.

In 1810, Risso, in his work on the Ichthyology
of Nice, laid the foundation of what has since been
termed "bathymetrical" distribution, or distribution in depth, by showing that regions of the sea
bottom of different depths could be distinguished
by the fishes which inhabit them.   There was the
littoral region between tide marks with its sand-
eels, pipe fishes, and blennies : the *seaweed region*,
extending from lowwater-mark to a depth of 450
feet, with its wrasses, rays, and flat fish ; and the
deep-sea region, from 450 feet to 1500 feet or more,
with its file-fish, sharks, gurnards, cod, and sword-
fish.

More than twenty years later, MM. Audouin and

Milne Edwards carried out the principle of distinguishing the Faunæ of different zones of depth much more minutely, in their " Recherches pour servir à l'Histoire Naturelle du Littoral de la France," published in 1832.

They divide the area included between high-water-mark and lowwater-mark of spring tides (which is very extensive, on account of the great rise and fall of the tide on the Normandy coast about St. Malo, where their observations were made) into four zones, each characterized by its peculiar invertebrate inhabitants. Beyond the fourth region they distinguish a fifth, which is never uncovered, and is inhabited by oysters, scallops, and large starfishes and other animals. Beyond this they seem to think that animal life is absent.[1]

Audouin and Milne Edwards were the first to see the importance of the bearing of a knowledge of the manner in which marine animals are distributed in depth, on geology. They suggest that, by this means, it will be possible to judge whether a fossiliferous stratum was formed upon the shore of an ancient sea, and even to determine whether it was deposited in shallower or deeper water on that shore; the association of shells of animals which live in different zones of depth will

---

[1] " Enfin plus bas encore, c'est-à-dire alors loin des côtes, le fond des eaux ne paraît plus être habité, du moins dans nos mers, par aucun de ces animaux" (l. c. tom. i. p. 237). The "ces animaux " leaves the meaning of the authors doubtful.

prove that the shells have been transported into the position in which they are found; while, on the other hand, the absence of shells in a deposit will not justify the conclusion that the waters in which it was formed were devoid of animal inhabitants, inasmuch as they might have been only too deep for habitation.

The new line of investigation thus opened by the French naturalists was followed up by the Norwegian, Sars, in 1835, by Edward Forbes, in our own country, in 1840,[1] and by Œrsted, in Denmark, a few years later. The genius of Forbes, combined with his extensive knowledge of botany, invertebrate zoology, and geology, enabled him to do more than any of his compeers, in bringing the importance of distribution in depth into notice; and his researches in the Ægean Sea,

---

[1] In the paper in the *Memoirs of the Survey* cited further on, Forbes writes :—

"In an essay 'On the Association of Mollusca on the British Coasts, considered with reference to Pleistocene Geology,' printed in [the *Edinburgh Academic Annual* for] 1840, I described the mollusca, as distributed on our shores and seas, in four great zones or regions, usually denominated 'The Littoral Zone,' 'The region of Laminariæ,' 'The region of Corallines,' and 'The region of Corals.' An extensive series of researches, chiefly conducted by the members of the committee appointed by the British Association to investigate the marine geology of Britain by means of the dredge, have not invalidated this classification, and the researches of Professor Lovén, in the Norwegian and Lapland seas, have borne out their correctness. The first two of the regions above mentioned had been previously noticed by Lamouroux, in his account of the distribution (vertically) of sea-weeds, by Audouin and Milne Edwards in their *Observations on the Natural History of the coast of France*, and by Sars in the preface to his *Beskrivelser og Jagttagelser*."

and still more his remarkable paper " On the Geological Relations of the existing Fauna and Flora of
the British Isles," published in 1846, in the first
volume of the " Memoirs of the Geological Survey
of Great Britain," attracted universal attention.

On the coasts of the British Islands, Forbes
distinguishes four zones or regions, the Littoral
(between tide marks), the Laminarian (between
lowwater-mark and 15 fathoms), the Coralline
(from 15 to 50 fathoms), and the Deep sea
or Coral region (from 50 fathoms to beyond 100
fathoms). But, in the deeper waters of the Ægean
Sea, between the shore and a depth of 300
fathoms, Forbes was able to make out no fewer
than eight zones of life, in the course of which the
number and variety of forms gradually diminished ;
until, beyond 300 fathoms, life disappeared altogether. Hence it appeared as if descent in the
sea had much the same effect on life, as ascent
on land. Recent investigations appear to show
that Forbes was right enough in his classification
of the facts of distribution in depth as they are to
be observed in the Ægean ; and though, at the
time he wrote, one or two observations were
extant which might have warned him not to
generalize too extensively from his Ægean experience, his own dredging work was so much
more extensive and systematic than that of any
other naturalist, that it is not wonderful he should
have felt justified in building upon it. Never-

theless, so far as the limit of the range of life in depth goes, Forbes' conclusion has been completely negatived, and the greatest depths yet attained show not even an approach to a "zero of life" :—

"During the several cruises of H.M. ships *Lightning* and *Porcupine* in the years 1868, 1869, and 1870," says Dr. Wyville Thomson, "fifty-seven hauls of the dredge were taken in the Atlantic at depths beyond 500 fathoms, and sixteen at depths beyond 1,000 fathoms, and, in all cases, life was abundant.   In 1869, we took two casts in depths greater than 2,000 fathoms. In both of these life was abundant ; and with the deepest cast, 2,435 fathoms, off the mouth of the Bay of Biscay, we took living, well-marked and characteristic examples of all the five invertebrate sub kingdoms.   And thus the question of the existence of abundant animal life at the bottom of the sea has been finally settled and for all depths, for there is no reason to suppose that the depth anywhere exceeds between three and four thousand fathoms ; and if there be nothing in the conditions of a depth of 2,500 fathoms to prevent the full develop-ment of a varied Fauna, it is impossible to suppose that even an additional thousand fathoms would make any great difference."[1]

As Dr. Wyville Thomson's recent letter, cited above, shows, the use of the trawl, at great depths, has brought to light a still greater diversity of life. Fishes came up from a depth of 600 to more than

[1] *The Depths of the Sea*, p. 30.   Results of a similar kind, obtained by previous observers, are stated at length in the sixth chapter, pp. 267–280.   The dredgings carried out by Count Pourtales, under the authority of Professor Peirce, the Super-intendent of the United States Coast Survey, in the years 1867, 1868, and 1869, are particularly noteworthy, and it is probably not too much to say, in the words of Professor Agassiz, "that we owe to the coast survey the first broad and comprehensive basis for an exploration of the sea bottom on a large scale, opening a new era in zoological and geological research."

1,000 fathoms, all "in a peculiar condition from the expansion of the air contained in their bodies. On their relief from the extreme pressure, their eyes, especially, had a singular appearance, protruding like great globes from their heads." Bivalve and univalve mollusca seem to be rare at the greatest depths; but starfishes, sea urchins, and other echinoderms, zoophytes, sponges, and protozoa abound.

It is obvious that the *Challenger* has the privilege of opening a new chapter in the history of the living world. She cannot send down her dredges and her trawls into these virgin depths of the great ocean without bringing up a discovery. Even though the thing itself may be neither "rich nor rare," the fact that it came from that depth, in that particular latitude and longitude, will be a new fact in distribution, and, as such, have a certain importance.

But it may be confidently assumed that the things brought up will very frequently be zoological novelties; or, better still, zoological antiquities, which, in the tranquil and little-changed depths of the ocean, have escaped the causes of destruction at work in the shallows, and represent the predominant population of a past age.

It has been seen that Audouin and Milne Edwards foresaw the general influence of the study of distribution in depth upon the interpretation of geological phenomena. Forbes connected the two orders of inquiry still more closely; and in the thoughtful essay " On the connection between the distribution of the existing Fauna and Flora of the British Isles, and the geological changes which have affected their area, especially during the epoch of the Northern drift," to which reference has already been made, he put forth a most pregnant suggestion.

In certain parts of the sea bottom in the immediate vicinity of the British Islands, as in the Clyde district, among the Hebrides, in the Moray Firth, and in the German Ocean, there are depressed areæ, forming a kind of submarine valleys, the centres of which are from 80 to 100 fathoms, or more, deep. These depressions are inhabited by assemblages of marine animals, which differ from those found over the adjacent and shallower region, and resemble those which are met with much farther north, on the Norwegian coast. Forbes called these Scandinavian detachments " Northern outliers."

How did these isolated patches of a northern population get into these deep places? To explain the mystery, Forbes called to mind the fact that, in the epoch which immediately preceded the present, the climate was much colder (whence the name of " glacial epoch " applied to it); and that the shells which are found fossil, or sub-fossil, in deposits of that age are precisely such as are now to be met with only in the Scandinavian, or still more Arctic, regions. Undoubtedly, during the glacial epoch, the general population of our seas had, universally, the northern aspect which is now presented only by the " northern outliers "; just as the vegetation of the land, down to the sea-level, had the northern character which is, at present, exhibited only by the plants which live on the tops of our mountains. But, as the glacial epoch passed away, and the present climatal conditions were developed, the northern plants were able to maintain themselves only on the bleak heights, on which southern forms could not compete with them. And, in like manner, Forbes suggested that, after the glacial epoch, the northern animals then inhabiting the sea became restricted to the deeps in which they could hold their own against invaders from the south, better fitted than they to flourish in the warmer waters of the shallows. Thus depth in the sea corresponded in its effect upon distribution to height on the land.

The same idea is applied to the explanation of a similar anomaly in the Fauna of the Ægean :—

"In the deepest of the regions of depth of the Ægean, the representation of a Northern Fauna is maintained, partly by identical and partly by representative forms. . . . The presence of the latter is essentially due to the law (of representation of parallels of latitude by zones of depth), whilst that of the former species depended on their transmission from their parent seas during a former epoch, and subsequent isolation. That epoch was doubtless the newer Pliocene or Glacial Era, when the *Mya truncata* and other northern forms now extinct in the Mediterranean, and found fossil in the Sicilian tertiaries, ranged into that sea. The changes which there destroyed the *shallow water* glacial forms, did not affect those living in the depths, and which still survive." [1]

The conception that the inhabitants of local depressions of the sea bottom might be a remnant of the ancient population of the area, which had held their own in these deep fastnesses against an invading Fauna, as Britons and Gaels have held out in Wales and in Scotland against encroaching Teutons, thus broached by Forbes, received a wider application than Forbes had dreamed of when the sounding machine first brought up specimens of the mud of the deep sea. As I have pointed out elsewhere,[2] it at once became obvious that the calcareous sticky mud of the Atlantic was made up, in the main, of shells of *Globigerina* and other *Foraminifera*, identical with those of which the true chalk is composed, and the identity extended even to the presence of those singular bodies, the Coccoliths and Coccospheres, the true nature of which is not yet made out. Here then were organisms, as old as the cretaceous epoch, still alive, and doing their work of rock-making at the bottom of existing seas. What if *Globigerina*

[1] *Memoirs of the Geological Survey of Great Britain*, Vol. i. p. 390.
[2] See above, "On a Piece of Chalk," p. 13,

and the Coccoliths should not be the only sur-
vivors of a world passed away, which are hidden
beneath three miles of salt water? The letter
which Dr. Wyville Thomson wrote to Dr. Carpenter in May, 1868, out of which all these expeditions have grown, shows that this query had
become a practical problem in Dr. Thomson's
mind at that time; and the desirableness of
solving the problem is put in the foreground of
his reasons for urging the Government to undertake the work of exploration :—

"Two years ago, M. Sars, Swedish Government Inspector of
Fisheries, had an opportunity, in his official capacity, of dredging off the Loffoten Islands at a depth of 300 fathoms. I
visited Norway shortly after his return, and had an opportunity
of studying with his father, Professor Sars, some of his results.
Animal forms were *abundant;* many of them were new to
science ; and among them was one of surpassing interest, the
small crinoid, of which you have a specimen, and which we at
once recognised as a degraded type of the *Apiocrinidæ,* an order
hitherto regarded as extinct, which attained its maximum in
the Pear Encrinites of the Jurassic period, and whose latest
representative hitherto known was the *Bourguettocrinus* of the
chalk. Some years previously, Mr. Absjornsen, dredging in 200
fathoms in the Hardangerfjord, procured several examples of a
Starfish (*Brisinga*), which seems to find its nearest ally in the
fossil genus *Protaster.* These observations place it beyond a
doubt that animal life is abundant in the ocean at depths
varying from 200 to 300 fathoms, that the forms at these great
depths differ greatly from those met with in ordinary dredgings,
and that, at all events in some cases, these animals are closely
allied to, and would seem to be directly descended from, the
Fauna of the early tertiaries.

"I think the latter result might almost have been anticipated ; and, probably, further investigation will largely add to this class of data, and will give us an opportunity of testing our determinations of the zoological position of some fossil types by an examination of the soft parts of their recent representatives. The main cause of the destruction, the migration, and the extreme modification of animal types, appear to be change of climate, chiefly depending upon oscillations of the earth's crust. These oscillations do not appear to have ranged, in the Northern portion of the Northern Hemisphere, much beyond 1,000 feet since the commencement of the Tertiary Epoch. The temperature of deep waters seems to be constant for all latitudes at 39° ; so that an immense area of the North Atlantic must have had its conditions unaffected by tertiary or post-tertiary oscillations." [1]

As we shall see, the assumption that the temperature of the deep sea is everywhere 39° F. (4° Cent.) is an error, which Dr. Wyville Thomson adopted from eminent physical writers; but the general justice of the reasoning is not affected by this circumstance, and Dr. Thomson's expectation has been, to some extent, already verified.

Thus besides *Globigerina,* there are eighteen species of deep-sea *Foraminifera* identical with species found in the chalk.   Imbedded in the chalky mud of the deep sea, in many localities, are innumerable cup-shaped sponges, provided with six-rayed silicious spicula, so disposed that the wall of the cup is formed of a lacework of flinty thread. Not less abundant, in some parts of the chalk formation, are the fossils known as *Ventriculites,* well described by

[1] *The Depths of the Sea,* pp. 51-52.

Dr. Thomson as "elegant vases or cups, with branching root-like bases, or groups of regularly or irregularly spreading tubes delicately fretted on the surface with an impressed network like the finest lace"; and he adds, "When we compare such recent forms as *Aphrocallistes, Iphiteon, Holtenia,* and *Askonema,* with certain series of the chalk *Ventriculites,* there cannot be the slightest doubt that they belong to the same family—in some cases to very nearly allied genera." [1]

Professor Duncan finds "several corals from the coast of Portugal more nearly allied to chalk forms than to any others."

The Stalked Crinoids or Feather Stars, so abundant in ancient times, are now exclusively confined to the deep sea, and the late explorations have yielded forms of old affinity, the existence of which has hitherto been unsuspected. The general character of the group of star fishes imbedded in the white chalk is almost the same as in the modern Fauna of the deep Atlantic. The sea urchins of the deep sea, while none of them are specifically identical with any chalk form, belong to the same general groups, and some closely approach extinct cretaceous genera.

Taking these facts in conjunction with the positive evidence of the existence, during the Cretaceous epoch, of a deep ocean where now lies the dry land of central and southern Europe,

[1] *The Depths of the Sea,* p. 484.

northern Africa, and western and southern Asia; and of the gradual diminution of this ocean during the older tertiary epoch, until it is represented at the present day by such teacupfuls as the Caspian, the Black Sea, and the Mediterranean; the supposition of Dr. Thomson and Dr. Carpenter that what is now the deep Atlantic, was the deep Atlantic (though merged in a vast easterly extension) in the Cretaceous epoch, and that the *Globigerina* mud has been accumulating there from that time to this, seems to me to have a great degree of probability. And I agree with Dr. Wyville Thomson against Sir Charles Lyell (it takes two of us to have any chance against his authority) in demurring to the assertion that "to talk of chalk having been uninterruptedly formed in the Atlantic is as inadmissible in a geographical as in a geological sense."

If the word "chalk" is to be used as a stratigraphical term and restricted to *Globigerina* mud deposited during the Cretaceous epoch, of course it is improper to call the precisely similar mud of more recent date, chalk. If, on the other hand, it is to be used as a mineralogical term, I do not see how the modern and the ancient chalks are to be separated—and, looking at the matter geographically, I see no reason to doubt that a boring rod driven from the surface of the mud which forms the floor of the mid-Atlantic would pass through one continuous mass of *Globigerina* mud, first of modern, then of tertiary, and then of mesozoic date; the "chalks" of different depths and ages being distinguished merely by the different forms of other organisms associated with the *Globigerinæ*.

On the other hand, I think it must be admitted that a belief in the continuity of the modern with the ancient chalk has nothing to do with the proposition that we can, in any sense whatever, be said to be still living in the Cretaceous epoch. When the *Challenger's* trawl brings up an *Ichthyosaurus*, along with a few living specimens of *Belemnites* and *Turrilites*, it may be admitted that she has come upon a cretaceous "outlier." A geological period is characterized not only by the presence of those creatures which lived in it, but by the absence of those which have only come into existence later; and, however large a proportion of true cretaceous forms may be discovered in the deep sea, the modern types associated with them must be abolished before the Fauna, as a whole, could, with any propriety, be termed Cretaceous.

I have now indicated some of the chief lines of Biological inquiry, in which the *Challenger* has special opportunities for doing good service, and in following which she will be carrying out the work already commenced by the *Lightning* and

Porcupine in their cruises of 1868 and subsequent years.

But biology, in the long run, rests upon physics, and the first condition for arriving at a sound theory of distribution in the deep sea, is the precise ascertainment of the conditions of life; or, in other words, a full knowledge of all those phenomena which are embraced under the head of the Physical Geography of the Ocean.

Excellent work has already been done in this direction, chiefly under the superintendence of Dr. Carpenter, by the *Lightning* and the *Porcupine*,[1] and some data of fundamental importance to the physical geography of the sea have been fixed beyond a doubt.

Thus, though it is true that sea-water steadily contracts as it cools down to its freezing point, instead of expanding before it reaches its freezing point as fresh water does, the truth has been steadily ignored by even the highest authorities in physical geography, and the erroneous conclusions deduced from their erroneous premises have been widely accepted as if they were ascertained facts. Of course, if sea-water, like fresh water, were heaviest at a temperature of 39° F. and got lighter as it approached 32° F., the water of the bottom of the deep sea could not be colder than 39°. But one of the first results of the careful ascertainment of the temperature

[1] *Proceedings of the Royal Society*, 1870 and 1872

at different depths, by means of thermometers specially contrived for the avoidance of the errors produced by pressure, was the proof that, below 1000 fathoms in the Atlantic, down to the greatest depths yet sounded, the water has a temperature always lower than 38° Fahr., whatever be the temperature of the water at the surface. And that this low temperature of the deepest water is probably the universal rule for the depths of the open ocean is shown, among others, by Captain Chimmo's recent observations in the Indian ocean, between Ceylon and Sumatra, where, the surface water ranging from 85°—81° Fahr., the temperature at the bottom, at a depth of 2270 to 2656 fathoms, was only from 34° to 32° Fahr.

As the mean temperature of the superficial layer of the crust of the earth may be taken at about 50° Fahr., it follows that the bottom layer of the deep sea in temperate and hot latitudes, is, on the average, much colder than either of the bodies with which it is in contact; for the temperature of the earth is constant, while that of the air rarely falls so low as that of the bottom water in the latitudes in question; and even when it does, has time to affect only a comparatively thin stratum of the surface water before the return of warm weather.

How does this apparently anomalous state of things come about? If we suppose the globe to be covered with a universal ocean, it can hardly be doubted that the cold of the regions towards the poles must tend to cause the superficial water of those regions to contract and become specifically heavier. Under these circumstances, it would have no alternative but to descend and spread over the sea bottom, while its place would be taken by warmer water drawn from the adjacent regions. Thus, deep, cold, polar-equatorial currents, and superficial, warmer, equatorial-polar currents, would be set up; and as the former would have a less velocity of rotation from west to east than the regions towards which they travel, they would not be due southerly or northerly currents, but south-westerly in the northern hemisphere, and north-westerly in the southern; while, by a parity of reasoning, the equatorial-polar warm currents would be north-easterly in the northern hemisphere, and south-easterly in the southern. Hence, as a north-easterly current has the same direction as a south-westerly wind, the direction of the northern equatorial-polar current in the extra-tropical part of its course would pretty nearly coincide with that of the anti-trade winds. The freezing of the surface of the polar sea would not interfere with the movement thus set up. For, however bad a conductor of heat ice may be, the unfrozen sea-water immediately in contact with the undersurface of the ice must needs be colder than that further off; and hence will constantly tend to descend through the subjacent warmer water.

In this way, it would seem inevitable that the surface waters of the northern and southern frigid zones must, sooner or later, find their way to the bottom of the rest of the ocean; and there accumulate to a thickness dependent on the rate at which they absorb heat from the crust of the earth below, and from the surface water above.

If this hypothesis be correct, it follows that, if any part of the ocean in warm latitudes is shut off from the influence of the cold polar underflow, the temperature of its deeps should be less cold than the temperature of corresponding depths in the open sea. Now, in the Mediterranean, Nature offers a remarkable experimental proof of just the kind needed. It is a landlocked sea which runs nearly east and west, between the twenty-ninth and forty-fifth parallels of north latitude. Roughly speaking, the average temperature of the air over it is 75° Fahr. in July and 48° in January.

This great expanse of water is divided by the peninsula of Italy (including Sicily), continuous with which is a submarine elevation carrying less than 1,200 feet of water, which extends from Sicily to Cape Bon in Africa, into two great pools —an eastern and a western. The eastern pool rapidly deepens to more than 12,000 feet, and sends off to the north its comparatively shallow branches, the Adriatic and the Ægean Seas. The western pool is less deep, though it reaches some 10,000 feet. And, just as the western end of the eastern pool communicates by a shallow passage, not a sixth of its greatest depth, with the western pool, so the western pool is separated from the Atlantic by a ridge which runs between Capes Trafalgar and Spartel, on which there is hardly 1,000 feet of water. All the water of the Mediterranean which lies deeper than about 150 fathoms, therefore, is shut off from that of the Atlantic, and there is no communication between the cold layer of the Atlantic (below 1,000 fathoms) and the Mediterranean. Under these circumstances, what is the temperature of the Mediterranean? Everywhere below 600 feet it is about 55° Fahr.; and consequently, at its greatest depths, it is some 20° warmer than the corresponding depths of the Atlantic.

It seems extremely difficult to account for this difference in any other way, than by adopting the views so strongly and ably advocated by Dr. Carpenter, that, in the existing distribution of land and water, such a circulation of the water of the ocean does actually occur, as theoretically must occur, in the universal ocean, with which we started.

It is quite another question, however, whether this theoretic circulation, true cause as it may be, is competent to give rise to such movements of sea-water, in mass, as those currents, which have commonly been regarded as northern extensions of the Gulf-stream. I shall not venture to touch

upon this complicated problem; but I may take
occasion to remark that the cause of a much
simpler phenomenon—the stream of Atlantic
water which sets through the Straits of Gibraltar,
eastward, at the rate of two or three miles an hour
or more, does not seem to be so clearly made out
as is desirable.

The facts appear to be that the water of the
Mediterranean is very slightly denser than that of
the Atlantic (1·0278 to 1·0265), and that the deep
water of the Mediterranean is slightly denser than
that of the surface; while the deep water of the
Atlantic is, if anything, lighter than that of the
surface. Moreover, while a rapid superficial current is setting in (always, save in exceptionally
violent easterly winds) through the Straits of
Gibraltar, from the Atlantic to the Mediterranean,
a deep undercurrent (together with variable side
currents) is setting out through the Straits, from
the Mediterranean to the Atlantic.

Dr. Carpenter adopts, without hesitation, the
view that the cause of this indraught of Atlantic
water is to be sought in the much more rapid
evaporation which takes place from the surface of
the Mediterranean than from that of the Atlantic;
and thus, by lowering the level of the former, gives
rise to an indraught from the latter.

But is there any sound foundation for the three
assumptions involved here? Firstly, that the
evaporation from the Mediterranean, as a whole,

is much greater than that from the Atlantic under corresponding parallels; secondly, that the rainfall over the Mediterranean makes up for evaporation less than it does over the Atlantic; and thirdly, supposing these two questions answered affirmatively: Are not these sources of loss in the Mediterranean fully covered by the prodigious quantity of fresh water which is poured into it by great rivers and submarine springs? Consider that the water of the Ebro, the Rhine, the Po, the Danube, the Don, the Dnieper, and the Nile, all flow directly or indirectly into the Mediterranean; that the volume of fresh water which they pour into it is so enormous that fresh water may sometimes be baled up from the surface of the sea off the Delta of the Nile, while the land is not yet in sight; that the water of the Black Sea is half fresh, and that a current of three or four miles an hour constantly streams from it Mediterraneanwards through the Bosphorus;—consider, in addition, that no fewer than ten submarine springs of fresh water are known to burst up in the Mediterranean, some of them so large that Admiral Smyth calls them " subterranean rivers of amazing volume and force "; and it would seem, on the face of the matter, that the sun must have enough to do to keep the level of the Mediterranean down; and that, possibly, we may have to seek for the cause of the small superiority in saline contents of the Mediterranean water in some condition other than solar evaporation.

Again, if the Gibraltar indraught is the effect of evaporation, why does it go on in winter as well as in summer ?

All these are questions more easily asked than answered ; but they must be answered before we can accept the Gibraltar stream as an example of a current produced by indraught with any comfort.

The Mediterranean is not included in the *Challenger's* route, but she will visit one of the most promising and little explored of hydrographical regions—the North Pacific, between Polynesia and the Asiatic and American shores; and doubtless the store of observations upon the currents of this region, which she will accumulate, when compared with what we know of the North Atlantic, will throw a powerful light upon the present obscurity of the Gulf-stream problem.

III

ON SOME OF THE RESULTS OF THE EXPEDITION OF H.M.S. *CHALLENGER*

[1875]

In May, 1873, I drew attention[1] to the important problems connected with the physics and natural history of the sea, to the solution of which there was every reason to hope the cruise of H.M.S. *Challenger* would furnish important contributions. The expectation then expressed has not been disappointed. Reports to the Admiralty, papers communicated to the Royal Society, and large collections which have already been sent home, have shown that the *Challenger's* staff have made admirable use of their great opportunities; and that, on the return of the expedition in 1874, their performance will be fully up to the level of their promise. Indeed, I am disposed to go so far as to say, that if nothing more came of the *Challenger's* expedition than

---

[1] See the preceding Essay.

has hitherto been yielded by her exploration of
the nature of the sea bottom at great depths, a
full scientific equivalent of the trouble and expense of her equipment would have been obtained.

In order to justify this assertion, and yet, at the
same time, not to claim more for Professor Wyville
Thomson and his colleagues than is their due, I
must give a brief history of the observations which
have preceded their exploration of this recondite
field of research, and endeavour to make clear
what was the state of knowledge in December,
1872, and what new facts have been added by the
scientific staff of the *Challenger*.   So far as I have
been able to discover, the first successful attempt
to bring up from great depths more of the sea
bottom than would adhere to a sounding-lead, was
made by Sir John Ross, in the voyage to the
Arctic regions which he undertook in 1818.   In
the Appendix to the narrative of that voyage,
there will be found an account of a very ingenious
apparatus called "clams"—a sort of double scoop
—of his own contrivance, which Sir John Ross
had made by the ship's armourer; and by which,
being in Baffin's Bay, in 72° 30′ N. and 77° 15′ W.,
he succeeded in bringing up from 1,050 fathoms
(or 6,300 feet), "several pounds" of a "fine green
mud," which formed the bottom of the sea in this
region.   Captain (now Sir Edward) Sabine, who
accompanied Sir John Ross on this cruise, says of
this mud that it was "soft and greenish, and that

the lead sunk several feet into it." A similar
"fine green mud" was found to compose the sea
bottom in Davis Straits by Goodsir in 1845.
Nothing is certainly known of the exact nature of
the mud thus obtained but we shall see that the
mud of the bottom of the Antarctic seas is described in curiously similar terms by Dr. Hooker,
and there is no doubt as to the composition of this
deposit.

In 1850, Captain Penny collected in Assistance
Bay, in Kingston Bay, and in Melville Bay,
which lie between 73° 45′ and 74° 40′ N., specimens of the residuum left by melted surface ice,
and of the sea bottom in these localities. Dr.
Dickie, of Aberdeen, sent these materials to
Ehrenberg, who made out[1] that the residuum of
the melted ice consisted for the most part of the
silicious cases of diatomaceous plants, and of the
silicious spicula of sponges; while, mixed with
these, were a certain number of the equally
silicious skeletons of those low animal organisms,
which were termed *Polycistineæ* by Ehrenberg, but
are now known as *Radiolaria*.

In 1856, a very remarkable addition to our
knowledge of the nature of the sea bottom in high
northern latitudes was made by Professor Bailey
of West Point. Lieutenant Brooke, of the United
States Navy, who was employed in surveying the

[1] *Ueber neue Anschauungen des kleinsten nördlichen Polar-
lebens.*—Monatsberichte d. K. Akad. Berlin, 1853.

Sea of Kamschatka, had succeeded in obtaining specimens of the sea bottom from greater depths than any hitherto reached, namely from 2,700 fathoms (16,200 feet) in 56° 46′ N., and 168° 18′ E.; and from 1,700 fathoms (10,200 feet) in 60° 15′ N· and 170° 53′ E. On examining these microscopically, Professor Bailey found, as Ehrenberg had done in the case of mud obtained on the opposite side of the Arctic region, that the fine mud was made up of shells of *Diatomacœ*, of spicula of sponges, and of *Radiolaria*, with a small admixture of mineral matters, but without a trace of any calcareous organisms.

Still more complete information has been obtained concerning the nature of the sea bottom in the cold zone around the south pole. Between the years 1839 and 1843, Sir James Clark Ross executed his famous Antarctic expedition, in the course of which he penetrated, at two widely distant points of the Antarctic zone, into the high latitudes of the shores of Victoria Land and of Graham's Land, and reached the parallel of 80° S. Sir James Ross was himself a naturalist of no mean acquirements, and Dr. Hooker,[1] the present President of the Royal Society, accompanied him as naturalist to the expedition, so that the observations upon the fauna and flora of the Antarctic regions made during this cruise were sure to have a peculiar value and importance, even had not the

[1] [Now Sir Joseph Hooker. 1894.]

attention of the voyagers been particularly directed to the importance of noting the occurrence of the minutest forms of animal and vegetable life in the ocean.

Among the scientific instructions for the voyage drawn up by a committee of the Royal Society, however, there is a remarkable letter from Von Humboldt to Lord Minto, then First Lord of the Admiralty, in which, among other things, he dwells upon the significance of the researches into the microscopic composition of rocks, and the discovery of the great share which microscopic organisms take in the formation of the crust of the earth at the present day, made by Ehrenberg in the years 1836-39. Ehrenberg, in fact, had shown that the extensive beds of "rotten-stone" or "Tripoli" which occur in various parts of the world, and notably at Bilin in Bohemia, consisted of accumulations of the silicious cases and skeletons of *Diatomaceæ*, sponges, and *Radiolaria*; he had proved that similar deposits were being formed by *Diatomaceæ*, in the pools of the Thiergarten in Berlin and elsewhere, and had pointed out that, if it were commercially worth while, rotten-stone might be manufactured by a process of diatom-culture. Observations conducted at Cuxhaven in 1839, had revealed the existence, at the surface of the waters of the Baltic, of living Diatoms and *Radiolaria* of the same species as those which, in a fossil state, constitute extensive rocks of tertiary age at Caltanisetta, Zante, and Oran, on the shores of the Mediterranean.

Moreover, in the fresh-water rotten-stone beds of Bilin, Ehrenberg had traced out the metamorphosis, effected apparently by the action of percolating water, of the primitively loose and friable deposit of organized particles, in which the silex exists in the hydrated or soluble condition. The silex, in fact, undergoes solution and slow redeposition, until, in ultimate result, the excessively fine-grained sand, each particle of which is a skeleton, becomes converted into a dense opaline stone, with only here and there an indication of an organism.

From the consideration of these facts, Ehrenberg, as early as the year 1839, had arrived at the conclusion that rocks, altogether similar to those which constitute a large part of the crust of the earth, must be forming, at the present day, at the bottom of the sea ; and he threw out the suggestion that even where no trace of organic structure is to be found in the older rocks, it may have been lost by metamorphosis.[1]

---

[1] *Ueber die noch jetzt zahlreich lebende Thierarten der Kreidebildung und den Organismus der Polythalamien. Abhandlungen der Kön. Akad. der Wissenchaften.* 1839. *Berlin.* 1841. I am afraid that this remarkable paper has been somewhat overlooked in the recent discussions of the relation of ancient rocks to modern deposits.

The results of the Antarctic exploration, as
stated by Dr. Hooker in the "Botany of the Antarctic Voyage," and in a paper which he read
before the British Association in 1847, are of the
greatest importance in connection with these
views, and they are so clearly stated in the former
work, which is somewhat inaccessible, that I make
no apology for quoting them at length—

"The waters and the ice of the South Polar Ocean were alike
found to abound with microscopic vegetables belonging to the
order *Diatomaceæ*. Though much too small to be discernible
by the naked eye, they occurred in such countless myriads as
to stain the berg and the pack ice wherever they were washed by
the swell of the sea ; and, when enclosed in the congealing
surface of the water, they imparted to the brash and pancake
ice a pale ochreous colour. In the open ocean, northward of
the frozen zone, this order, though no doubt almost universally
present, generally eludes the search of the naturalist; except
when its species are congregated amongst that mucous scum
which is sometimes seen floating on the waves, and of whose
real nature we are ignorant ; or when the coloured contents of
the marine animals who feed on these Algæ are examined. To
the south, however, of the belt of ice which encircles the globe,
between the parallels of 50° and 70° S., and in the waters comprised between that belt and the highest latitude ever attained by
man, this vegetation is very conspicuous, from the contrast
between its colour and the white snow and ice in which it is
imbedded. Insomuch, that in the eightieth degree, all the
surface ice carried along by the currents, the sides of every
berg, and the base of the great Victoria Barrier itself, within
reach of the swell, were tinged brown, as if the polar waters
were charged with oxide of iron.

"As the majority of these plants consist of very simple vegetable cells, enclosed in indestructible silex (as other Algæ are in
carbonate of lime), it is obvious that the death and decomposi-

tion of such multitudes must form sedimentary deposits, propor-
tionate in their extent to the length and exposure of the coast
against which they are washed, in thickness to the power of
such agents as the winds, currents, and sea, which sweep
them more energetically to certain positions, and in purity, to
the depth of the water and nature of the bottom. Hence we
detected their remains along every icebound shore, in the depths
of the adjacent ocean, between 80 and 400 fathoms. Off
Victoria Barrier (a perpendicular wall of ice between one and
two hundred feet above the level of the sea) the bottom of the
ocean was covered with a stratum of pure white or green mud,
composed principally of the silicious shells of the *Diatomaceœ*.
These, on being put into water, rendered it cloudy like milk,
and took many hours to subside. In the very deep water off
Victoria and Graham's Land, this mud was particularly pure and
fine ; but towards the shallow shores there existed a greater or
less admixture of disintegrated rock and sand ; so that the
organic compounds of the bottom frequently bore but a small
proportion to the inorganic." . . .

" The universal existence of such an invisible vegetation as
that of the Antarctic Ocean, is a truly wonderful fact, and the
more from its not being accompanied by plants of a high order.
During the years we spent there, I had been accustomed to
regard the phenomena of life as differing totally from what obtains
throughout all other latitudes, for everything living appeared
to be of animal origin. The ocean swarmed with *Mollusca*, and
particularly entomostracous *Crustacea*, small whales, and por-
poises ; the sea abounded with penguins and seals, and the air
with birds ; the animal kingdom was ever present, the larger
creatures preying on the smaller, and these again on smaller
still ; all seemed carnivorous. The herbivorous were not recognised, because feeding on a microscopic herbage, of whose true
nature I had formed an erroneous impression. It is, therefore,
with no little satisfaction that I now class the *Diatomaceœ* with
plants, probably maintaining in the South Polar Ocean that
balance between the vegetable and the animal kingdoms which
prevails over the surface of our globe. Nor is the sustenance
and nutrition of the animal kingdom the only function these

minute productions may perform ; they may also be the purifiers
of the vitiated atmosphere, and thus execute in the Antarctic
latitudes the office of our trees and grass turf in the temperate
regions, and the broad leaves of the palm, &c., in the
tropics." . . . .

With respect to the distribution of the
Diatomaceæ, Dr. Hooker remarks :—

" There is probably no latitude between that of Spitzbergen
and Victoria Land, where some of the species of either country
do not exist : Iceland, Britain, the Mediterranean Sea, North and
South America, and the South Sea Islands, all possess Antarctic
Diatomaceæ. The silicious coats of species only known living
in the waters of the South Polar Ocean, have, during past
ages, contributed to the formation of rocks ; and thus they out-
live several successive creations of organized beings. The
phonolite stones of the Rhine, and the Tripoli stone, contain
species identical with what are now contributing to form a sedimentary deposit (and perhaps, at some future period, a bed of
rock) extending in one continuous stratum for 400 measured
miles. I allude to the shores of the Victoria Barrier, along
whose coast the soundings examined were invariably charged
with diatomaceous remains, constituting a bank which stretches
200 miles north from the base of Victoria Barrier, while the
average depth of water above it is 300 fathoms, or 1,800 feet.
Again, some of the Antarctic species have been detected floating
in the atmosphere which overhangs the wide ocean between
Africa and America. The knowledge of this marvellous fact we
owe to Mr. Darwin, who, when he was at sea off the Cape de
Verd Islands, collected an impalpable powder which fell on
Captain Fitzroy's ship. He transmitted this dust to Ehrenberg,
who ascertained it to consist of the silicious coats, chiefly of
American *Diatomaceæ*, which were being wafted through the
upper region of the air, when some meteorological phenomena
checked them in their course and deposited them on the ship
and surface of the ocean.

" The existence of the remains of many species of this order (and amongst them some Antarctic ones) in the volcanic ashes, pumice, and scoriæ of active and extinct volcanoes (those of the Mediterranean Sea and Ascension Island, for instance) is a fact bearing immediately upon the present subject. Mount Erebus, a volcano 12,400 feet high, of the first class in dimensions and energetic action, rises at once from the ocean in the seventy-eighth degree of south latitude, and abreast of the *Diatomaceæ* bank, which reposes in part on its base. Hence it may not appear preposterous to conclude that, as Vesuvius receives the waters of the Mediterranean, with its fish, to eject them by its crater, so the subterranean and subaqueous forces which maintain Mount Erebus in activity may occasionally receive organic matter from the bank, and disgorge it, together with those volcanic products, ashes and pumice.

"Along the shores of Graham's Land and the South Shetland Islands, we have a parallel combination of igneous and aqueous action, accompanied with an equally copious supply of *Diatom-aceæ*. In the Gulf of Erebus and Terror, fifteen degrees north of Victoria Land, and placed on the opposite side of the globe, the soundings were of a similar nature with those of the Victoria Land and Barrier, and the sea and ice as full of *Diatomaceæ*. This was not only proved by the deep sea lead, but by the examination of bergs which, once stranded, had floated off and become reversed, exposing an accumulation of white friable mud frozen to their bases, which abounded with these vegetable remains."

The *Challenger* has explored the Antarctic seas in a region intermediate between those examined by Sir James Ross's expedition; and the observations made by Dr. Wyville Thomson and his colleagues in every respect confirm those of Dr. Hooker :—

"On the 11th of February, lat. 60° 52′ S., long. 80° 20′ E., and March 3, lat. 53° 55′ S., long. 108° 35′ E., the sounding

instrument came up filled with a very fine cream-coloured paste,
which scarcely effervesced with acid, and dried into a very light,
impalpable, white powder. This, when examined under the
microscope, was found to consist almost entirely of the frustules
of Diatoms, some of them wonderfully perfect in all the details
of their ornament, and many of them broken up. The species
of Diatoms entering into this deposit have not yet been worked
up, but they appear to be referable chiefly to the genera *Fragil-
laria, Coscinodiscus, Chætoceros, Asteromphalus,* and *Dictyocha,*
with fragments of the separated rods of a singular silicious
organism, with which we were unacquainted, and which made
up a large proportion of the finer matter of this deposit. Mixed
with the Diatoms there were a few small *Globigerinæ,* some of the
tests and spicules of Radiolarians, and some sand particles ; but
these foreign bodies were in too small proportion to affect the
formation as consisting practically of Diatoms alone. On the
4th of February, in lat. 52°, 29′ S., long., 71° 36′ E., a little to
the north of the Heard Islands, the tow-net, dragging a few
fathoms below the surface, came up nearly filled with a pale yellow
gelatinous mass. This was found to consist entirely of Diatoms
of the same species as those found at the bottom. By far the
most abundant was the little bundle of silicious rods, fastened
together loosely at one end, separating from one another at the
other end, and the whole bundle loosely twisted into a spindle.
The rods are hollow, and contain the characteristic endochrome
of the *Diatomaceæ.* Like the *Globigerina* ooze, then, which it
succeeds to the southward in a band apparently of no great
width, the materials of this silicious deposit are derived entirely
from the surface and intermediate depths. It is somewhat
singular that Diatoms did not appear to be in such large numbers on the surface over the Diatom ooze as they were a little
further north. This may perhaps be accounted for by our not
having struck their belt of depth with the tow-net ; or it is
possible that when we found it on the 11th of February the bottom
deposit was really shifted a little to the south by the warm
current, the excessively fine flocculent *débris* of the Diatoms
taking a certain time to sink. The belt of Diatom ooze is
certainly a little further to the southward in long. 83° E., in

the path of the reflux of the Agulhas current, than in long. 108° E.

"All along the edge of the ice-pack—everywhere, in fact, to the south of the two stations—on the 11th of February on our southward voyage, and on the 3rd of March on our return, we brought up fine sand and grayish mud, with small pebbles of quartz and felspar, and small fragments of mica-slate, chlorite-slate, clay-slate, gneiss, and granite. This deposit, I have no doubt, was derived from the surface like the others, but in this case by the melting of icebergs and the precipitation of foreign matter contained in the ice.

"We never saw any trace of gravel or sand, or any material necessarily derived from land, on an iceberg. Several showed vertical or irregular fissures filled with discoloured ice or snow ; but, when looked at closely, the discoloration proved usually to be very slight, and the effect at a distance was usually due to the foreign material filling the fissure reflecting light less perfectly than the general surface of the berg. I conceive that the upper surface of one of these great tabular southern ice-bergs, including by far the greater part of its bulk, and culmin-ating in the portion exposed above the surface of the sea, was formed by the piling up of successive layers of snow during the period, amounting perhaps to several centuries, during which the ice-cap was slowly forcing itself over the low land and out to sea over a long extent of gentle slope, until it reached a depth considerably above 200 fathoms, when the lower specific weight of the ice caused an upward strain which at length overcame the cohesion of the mass, and portions were rent off and floated away. If this be the true history of the formation of these icebergs, the absence of all land *débris* in the portion exposed above the surface of the sea is readily understood. If any such exist, it must be confined to the lower part of the berg, to that part which has at one time or other moved on the floor of the ice-cap.

"The icebergs, when they are first dispersed, float in from 200 to 250 fathoms. When, therefore, they have been drifted to latitudes of 65° or 64° S., the bottom of the berg just reaches the layer at which the temperature of the water is distinctly rising, and it is rapidly melted, and the mud and pebbles with which it is more or less charged are precipitated. That this precipitation takes place all over the area where the icebergs are breaking up, constantly, and to a considerable extent, is evident from the fact of the soundings being entirely composed of such deposits ; for the Diatoms, *Globigerinæ*, and radiolarians are present on the surface in large numbers ; and unless the deposit from the ice were abundant it would soon be covered and masked by a layer of the exuvia of surface organisms."

The observations which have been detailed leave no doubt that the Antarctic sea bottom, from a little to the south of the fiftieth parallel, as far as 80° S., is being covered by a fine deposit of silicious mud, more or less mixed, in some parts, with the ice-borne *débris* of polar lands and with the ejections of volcanoes. The silicious particles which constitute this mud, are derived, in part, from the diatomaceous plants and radiolarian animals which throng the surface, and, in part, from the spicula of sponges which live at the bottom. The evidence respecting the corresponding Arctic area is less complete, but it is sufficient to justify the conclusion that an essentially similar silicious cap is being formed around the northern pole.

There is no doubt that the constituent particles of this mud may agglomerate into a dense rock, such as that formed at Oran, on the shores of the Mediterranean, which is made up of similar materials. Moreover, in the case of freshwater deposits of this kind, it is certain that the action of percolating water may convert the originally soft and friable, fine-grained sandstone into a dense, semi-transparent opaline stone, the silicious organized skeletons being dissolved, and the silex re-deposited in an amorphous state. Whether such a metamorphosis as this occurs in submarine deposits, as well as in those formed in fresh water, does not appear; but there seems no reason to doubt that it may. And hence it may not be hazardous to conclude that very ordinary meta-morphic agencies may convert these polar caps into a form of quartzite.

In the great intermediate zone, occupying some 110° of latitude, which separates the circumpolar Arctic and Antarctic areas of silicious deposit, the Diatoms and *Radiolaria* of the surface water and the sponges of the bottom do not die out, and, so far as some forms are concerned, do not even appear to diminish in total number; though, on a rough estimate, it would appear that the proportion of *Radiolaria* to Diatoms is much greater than in the colder seas. Nevertheless the composition of the deep-sea mud of this intermediate zone is entirely different from that of the circumpolar regions.

The first exact information respecting the nature of this mud at depths greater than 1,000 fathoms was given by Ehrenberg, in the account which he published in the "Monatsberichte" of the Berlin Academy for the year 1853, of the soundings obtained by Lieut. Berryman, of the United States Navy, in the North Atlantic, between Newfoundland and the Azores.

Observations which confirm those of Ehrenberg in all essential respects have been made by Professor Bailey, myself, Dr. Wallich, Dr. Carpenter, and Professor Wyville Thomson, in their earlier cruises; and the continuation of the *Globigerina* ooze over the South Pacific has been proved by the recent work of the *Challenger*, by which it is also shown, for the first time, that, in passing from the equator to high southern latitudes, the number and variety of the *Foraminifera* diminishes, and even the *Globigerinæ* become dwarfed. And this result, it will be observed, is in entire accordance with the fact already mentioned that, in the sea of Kamschatka, the deepsea mud was found by Bailey to contain no calcareous organisms.

Thus, in the whole of the "intermediate zone," the silicious deposit which is being formed there, as elsewhere, by the accumulation of spongespicula, *Radiolaria*, and Diatoms, is obscured and overpowered by the immensely greater amount of calcareous sediment, which arises from the aggregation of the skeletons of dead *Foraminifera*. The similarity of the deposit, thus composed of a large percentage of carbonate of lime, and a small percentage of silex, to chalk, regarded merely as a

kind of rock, which was first pointed out by
Ehrenberg,[1] is now admitted on all hands; nor
can it be reasonably doubted, that ordinary meta-
morphic agencies are competent to convert the
"modern chalk" into hard limestone or even into
crystalline marble.

Ehrenberg appears to have taken it for granted
that the *Globigerinæ* and other *Foraminifera* which
are found in the deep-sea mud, live at the great
depths in which their remains are found; and he
supports this opinion by producing evidence that
the soft parts of these organisms are preserved,
and may be demonstrated by removing the calcareous matter with dilute acids.   In 1857, the

[1] The following passages in Ehrenberg's memoir on *The
Organisms in the Chalk which are still living* (1839), are conclusive :—

"7. The dawning period of the existing living organic creation,
if such a period is distinguishable (which is doubtful), can only
be supposed to have existed on the other side of, and below, the
chalk formation ; and thus, either the chalk, with its wide-
spread and thick beds, must enter into the series of newer
formations ; or some of the accepted four great geological periods,
the quaternary, tertiary, and secondary formations, contain
organisms which still live.   It is more probable, in the proportion of 3 to 1, that the transition or primary period is not
different, but that it is only more difficult to examine and
understand, by reason of the gradual and prolonged chemical
decomposition and metamorphosis of many of its organic
constituents."

"10. By the mass-forming *Infusoria* and *Polythalamia*,
secondary are not distinguishable from tertiary formations ; and,
from what has been said, it is possible that, at this very day,
rock masses are forming in the sea, and being raised by volcanic
agencies, the constitution of which, on the whole, is altogether
similar to that of the chalk.   The chalk remains distinguishable
by its organic remains as a formation, but not as a kind of
rock."

evidence for and against this conclusion appeared
to me to be insufficient to warrant a positive conclusion one way or the other, and I expressed
myself in my report to the Admiralty on Captain
Dayman's soundings in the following terms :—

"When we consider the immense area over which this
deposit is spread, the depth at which its formation is going on,
and its similarity to chalk, and still more to such rocks as the
marls of Caltanisetta, the question, whence are all these organisms derived ? becomes one of high scientific interest.

"Three answers have suggested themselves :—

"In accordance with the prevalent view of the limitation of
life to comparatively small depths, it is imagined either : 1, that
these organisms have drifted into their present position from
shallower waters ; or 2, that they habitually live at the surface
of the ocean, and only fall down into their present position.

"1. I conceive that the first supposition is negatived by the
extremely marked zoological peculiarity of the deep-sea fauna.

"Had the *Globigerinæ* been drifted into their present position
from shallow water, we should find a very large proportion of
the characteristic inhabitants of shallow waters mixed with
them, and this would the more certainly be the case, as the
large *Globigerinæ*, so abundant in the deep-sea soundings, are,
in proportion to their size, more solid and massive than almost
any other *Foraminifera*.   But the fact is that the proportion of
other *Foraminifera* is exceedingly small, nor have I found as
yet, in the deep-sea deposits, any such matters as fragments
of molluscous shells, of *Echini*, &c., which abound in shallow
waters, and are quite as likely to be drifted as the heavy *Globigerinæ*.   Again, the relative proportions of young and fully
formed *Globigerinæ* seem inconsistent with the notion that they
have travelled far.   And it seems difficult to imagine why, had
the deposit been accumulated in this way, *Coscinodisci* should
so almost entirely represent the *Diatomaceæ*.

"2. The second hypothesis is far more feasible, and is
strongly supported by the fact that many *Polycistineæ* [*Radiola-*

ria] and *Coscinodisci* are well known to live at the surface of the ocean. Mr. Macdonald, Assistant-Surgeon of H.M.S. *Herald*, now in the South-Western Pacific, has lately sent home some very valuable observations on living forms of this kind, met with in the stomachs of oceanic mollusks, and therefore certainly inhabitants of the superficial layer of the ocean. But it is a singular circumstance that only one of the forms figured by Mr. Macdonald is at all like a *Globigerina*, and there are some peculiarities about even this which make me greatly doubt its affinity with that genus. The form, indeed, is not unlike that of a *Globigerina*, but it is provided with long radiating processes, of which I have never seen any trace in *Globigerina*. Did they exist, they might explain what otherwise is a great objection to this view, viz., how is it conceivable that the heavy *Globigerina* should maintain itself at the surface of the water ?

" If the organic bodies in the deep-sea soundings have neither been drifted, nor have fallen from above, there remains but one alternative—they must have lived and died where they are.

" Important objections, however, at once suggest themselves to this view. How can animal life be conceived to exist under such conditions of light, temperature, pressure, and aeration as must obtain at these vast depths ?

" To this one can only reply that we know for a certainty that even very highly-organized animals do continue to live at a depth of 300 and 400 fathoms, inasmuch as they have been dredged up thence ; and that the difference in the amount of light and heat at 400 and at 2,000 fathoms is probably, so to speak, very far less than the difference in complexity of organisation between these animals and the humbler *Protozoa* and *Protophyta* of the deep-sea soundings.

" I confess, though as yet far from regarding it proved that the *Globigerinæ* live at these depths, the balance of probabilities seems to me to incline in that direction. And there is one circumstance which weighs strongly in my mind. It may be taken as a law that any genus of animals which is found far back in time is capable of living under a great variety of circumstances as regards light, temperature, and pressure. Now, the genus *Globigerina* is abundantly represented in the cretaceous epoch, and perhaps earlier.

"I abstain, however, at present from drawing any positive conclusions, preferring rather to await the result of more extended observations." [1]

Dr. Wallich, Professor Wyville Thomson, and Dr. Carpenter concluded that the *Globigerinæ* live at the bottom. Dr. Wallich writes in 1862—" By sinking very fine gauze nets to considerable depths, I have repeatedly satisfied myself that *Globigerina* does not occur in the superficial strata of the ocean." [2] Moreover, having obtained certain living star-fish from a depth of 1,260 fathoms, and found their stomachs full of " fresh-looking *Globigerinæ* " and their *débris*—he adduces this fact in support of his belief that the *Globigerinæ* live at the bottom.

On the other hand, Müller, Haeckel, Major Owen, Mr. Gwyn Jeffries, and other observers, found that *Globigerinæ*, with the allied genera *Orbulina* and *Pulvinulina*, sometimes occur abundantly at the surface of the sea, the shells of these pelagic forms being not unfrequently provided with the long spines noticed by Macdonald ; and in 1865 and 1866, Major Owen more especially insisted on the importance of this fact. The recent work of the *Challenger* fully confirms Major Owen's statement. In the paper recently pub-

[1] Appendix to Report on Deep-sea Soundings in the Atlantic Ocean, by Lieut.-Commander Joseph Dayman. 1857.
[2] The *North Atlantic Sea-bed*, p. 137.

lished in the proceedings of the Royal Society,[1] from which a quotation has already been made, Professor Wyville Thomson says :—

"I had formed and expressed a very strong opinion on the matter. It seemed to me that the evidence was conclusive that the *Foraminifera* which formed the *Globigerina* ooze lived on the bottom, and that the occurrence of individuals on the surface was accidental and exceptional ; but after going into the thing carefully, and considering the mass of evidence which has been accumulated by Mr. Murray, I now admit that I was in error ; and I agree with him that it may be taken as proved that all the materials of such deposits, with the exception, of course, of the remains of animals which we now know to live at the bottom at all depths, which occur in the deposit as foreign bodies, are derived from the surface.

"Mr. Murray has combined with a careful examination of the soundings a constant use of the tow-net, usually at the surface, but also at depths of from ten to one hundred fathoms ; and he finds the closest relation to exist between the surface fauna of any particular locality and the deposit which is taking place at the bottom. In all seas, from the equator to the polar ice, the tow-net contains *Globigerinæ*. They are more abundant and of a larger size in warmer seas ; several varieties, attaining a large size and presenting marked varietal characters, are found in the intertropical area of the Atlantic. In the latitude of Kerguelen they are less numerous and smaller, while further south they are still more dwarfed, and only one variety, the typical *Globigerina bulloides*, is represented. The living *Globigerinæ* from the tow-net are singularly different in appearance from the dead shells we find at the bottom. The shell is clear and transparent, and each of the pores which penetrate it is surrounded by a raised crest, the crest round adjacent pores coalescing into a roughly

---

[1] "Preliminary Notes on the Nature of the Sea-bottom procured by the soundings of H.M.S. *Challenger* during her cruise in the Southern Seas, in the early part of the year 1874."— *Proceedings of the Royal Society*, Nov. 26, 1874.

hexagonal network, so that the pores appear to lie at the
bottom of a hexagonal pit. At each angle of this hexagon the
crest gives off a delicate flexible calcareous spine, which is sometimes four or five times the diameter of the shell in length.
The spines radiate symmetrically from the direction of the
centre of each chamber of the shell, and the sheaves of long
transparent needles crossing one another in different directions
have a very beautiful effect. The smaller inner chambers of the
shell are entirely filled with an orange-yellow granular sarcode ;
and the large terminal chamber usually contains only a small
irregular mass, or two or three small masses run together, of
the same yellow sarcode stuck against one side, the remainder of
the chamber being empty. No definite arrangement and no
approach to structure was observed in the sarcode, and no
differentiation, with the exception of round bright-yellow oil-
globules, very much like those found in some of the radiolarians,
which are scattered, apparently irregularly, in the sarcode. We
never have been able to detect, in any of the large number of
Globigerinæ which we have examined, the least trace of pseudo-
podia, or any extension, in any form, of the sarcode beyond the
shell.

*   *   *   *   *

" In specimens taken with the tow-net the spines are very
usually absent ; but that is probably on account of their extreme
tenuity ; they are broken off by the slightest touch. In fresh
examples from the surface, the dots indicating the origin of the
lost spines may almost always be made out with a high power.
There are never spines on the *Globigerinæ* from the bottom,
even in the shallowest water."

There can now be no doubt, therefore, that
Globigerinæ live at the top of the sea ; but the
question may still be raised whether they do not
also live at the bottom. In favour of this view, it
has been urged that the shells of the *Globigerinæ*
of the surface never possess such thick walls as

those which are found at the bottom, but I confess
that I doubt the accuracy of this statement.
Again, the occurrence of minute *Globigerinæ* in all
stages of development, at the greatest depths, is
brought forward as evidence that they live *in situ.*
But considering the extent to which the surface
organisms are devoured, without discrimination of
young and old, by *Salpæ* and the like, it is not
wonderful that shells of all ages should be among
the rejectamenta. Nor can the presence of the
soft parts of the body in the shells which form
the *Globigerina* ooze, and the fact, if it be one,
that animals living at the bottom use them as
food, be considered as conclusive evidence that
the *Globigerinæ* live at the bottom. Such as die
at the surface, and even many of those which are
swallowed by other animals, may retain much of
their protoplasmic matter when they reach the
depths at which the temperature sinks to 34° or
32° Fahrenheit, where decomposition must become
exceedingly slow.

Another consideration appears to me to be in
favour of the view that the *Globigerinæ* and their
allies are essentially surface animals. This is the
fact brought out by the *Challenger's* work, that
they have a southern limit of distribution, which
can hardly depend upon anything but the temperature of the surface water. And it is to
be remarked that this southern limit occurs at a
lower latitude in the Antarctic seas than it does

in the North Atlantic. According to Dr. Wallich
("The North Atlantic Sea Bed," p. 157) *Globigerina* is the prevailing form in the deposits
between the Faroe Islands and Iceland, and between Iceland and East Greenland—or, in other
words, in a region of the sea-bottom which lies
altogether north of the parallel of 60° N.; while
in the southern seas, the *Globigerinæ* become
dwarfed and almost disappear between 50° and
55° S. On the other hand, in the sea of
Kamschatka, the *Globigerinæ* have vanished in
56° N., so that the persistence of the *Globigerina*
ooze in high latitudes, in the North Atlantic,
would seem to depend on the northward curve of
the isothermals peculiar to this region; and it is
difficult to understand how the formation of
Globigerina ooze can be affected by this climatal
peculiarity unless it be effected by surface animals.

Whatever may be the mode of life of the
Foraminifera, to which the calcareous element of
the deep-sea "chalk" owes its existence, the fact
that it is the chief and most widely spread
material of the sea-bottom in the intermediate
zone, throughout both the Atlantic and Pacific
Oceans, and the Indian Ocean, at depths from a
few hundred to over two thousand fathoms, is
established. But it is not the only extensive
deposit which is now taking place. In 1853,
Count Pourtalès, an officer of the United States
Coast Survey, which has done so much for

scientific hydrography, observed, that the mud
forming the sea-bottom at depths of one hundred
and fifty fathoms, in 31° 32′ N., 79° 35′ W., off
the Coast of Florida, was " a mixture, in about
equal proportions, of *Globigerinœ* and black sand,
probably greensand, as it makes a green mark
when crushed on paper." Professor Bailey,
examining these grains microscopically, found
that they were casts of the interior cavities of
Foraminifera, consisting of a mineral known as
Glauconite, which is a silicate of iron and alumina.
In these casts the minutest cavities and finest
tubes in the Foraminifer were sometimes repro-
duced in solid counterparts of the glassy mineral,
while the calcareous original had been entirely
dissolved away.

Contemporaneously with these observations,
the indefatigable Ehrenberg had discovered that
the " greensands " of the geologist were largely
made up of casts of a similar character, and proved
the existence of *Foraminifera* at a very ancient
geological epoch, by discovering such casts in a
greensand of Lower Silurian age, which occurs
near St. Petersburg.

Subsequently, Messrs. Parker and Jones discovered similar casts in process of formation, the
original shell not having disappeared, in specimens
of the sea-bottom of the Australian seas, brought
home by the late Professor Jukes. And the
Challenger has observed a deposit of a similar

character in the course of the Agulhas current, near the Cape of Good Hope, and in some other localities not yet defined.

It would appear that this infiltration of *Foraminifera* shells with *Glauconite* does not take place at great depths, but rather in what may be termed a sublittoral region, ranging from a hundred to three hundred fathoms. It cannot be ascribed to any local cause, for it takes place, not only over large areas in the Gulf of Mexico and the Coast of Florida, but in the South Atlantic and in the Pacific. But what are the conditions which determine its occurrence, and whence the silex, the iron, and the alumina (with perhaps potash and some other ingredients in small quantity) of which the *Glauconite* is composed, proceed, is a point on which no light has yet been thrown. For the present we must be content with the fact that, in certain areas of the " intermediate zone," greensand is replacing and representing the primitively calcareo-silicious ooze.

The investigation of the deposits which are now being formed in the basin of the Mediterranean, by the late Professor Edward Forbes, by Professor Williamson, and more recently by Dr. Carpenter, and a comparison of the results thus obtained with what is known of the surface fauna, have brought to light the remarkable fact, that while the surface and the shallows abound with

Foraminifera and other calcareous shelled organisms, the indications of life become scanty at depths beyond 500 or 600 fathoms, while almost all traces of it disappear at greater depths, and at 1,000 to 2,000 fathoms the bottom is covered with a fine clay.

Dr. Carpenter has discussed the significance of this remarkable fact, and he is disposed to attribute the absence of life at great depths, partly to the absence of any circulation of the water of the Mediterranean at such depths, and partly to the exhaustion of the oxygen of the water by the organic matter contained in the fine clay, which he conceives to be formed by the finest particles of the mud brought down by the rivers which flow into the Mediterranean.

However this may be, the explanation thus offered of the presence of the fine mud, and of the absence of organisms which ordinarily live at the bottom, does not account for the absence of the skeletons of the organisms which undoubtedly abound at the surface of the Mediterranean; and it would seem to have no application to the remarkable fact discovered by the *Challenger*, that in the open Atlantic and Pacific Oceans, in the midst of the great intermediate zone, and thousands of miles away from the embouchure of any river, the sea-bottom, at depths approaching to and beyond 3,000 fathoms, no longer consists of *Globigerina* ooze, but of an excessively fine red clay.

Professor Thomson gives the following account of this capital discovery :—

"According to our present experience, the deposit of *Globigerina* ooze is limited to water of a certain depth, the extreme limit of the pure characteristic formation being placed at a depth of somewhere about 2,250 fathoms. Crossing from these shallower regions occupied by the ooze into deeper soundings, we find, universally, that the calcareous formation gradually passes into, and is finally replaced by, an extremely fine pure clay, which occupies, speaking generally, all depths below 2,500 fathoms, and consists almost entirely of a silicate of the red oxide of iron and alumina. The transition is very slow, and extends over several hundred fathoms of increasing depth ; the shells gradually lose their sharpness of outline, and assume a kind of 'rotten' look and a brownish colour, and become more and more mixed with a fine amorphous red-brown powder, which increases steadily in proportion until the lime has almost entirely disappeared. This brown matter is in the finest possible state of subdivision, so fine that when, after sifting it to separate any organisms it might contain, we put it into jars to settle, it remained for days in suspension, giving the water very much the appearance and colour of chocolate.

"In indicating the nature of the bottom on the charts, we came, from experience and without any theoretical considerations, to use three terms for soundings in deep water. Two of these, Gl. oz. and r. cl., were very definite, and indicated strongly-marked formations, with apparently but few characters in common ; but we frequently got soundings which we could not exactly call '*Globigerina* ooze' or red clay,' and before we were fully aware of the nature of these, we were in the habit of indicating them as 'grey ooze' (gr. oz.) We now recognise the 'grey ooze' as an intermediate stage between the *Globigerina* ooze and the red clay ; we find that on one side, as it were, of an ideal line, the red clay contains more and more of the material of the calcareous ooze, while on the other, the ooze is mixed with an increasing proportion of 'red clay.'

"Although we have met with the same phenomenon so frequently, that we were at length able to predict the nature of the bottom from the depth of the soundings with absolute certainty for the Atlantic and the Southern Sea, we had, perhaps, the best opportunity of observing it in our first section across the Atlantic, between Teneriffe and St. Thomas. The first four stations on this section, at depths from 1,525 to 2,220 fathoms, show *Globigerina* ooze. From the last of these, which is about 300 miles from Teneriffe, the depth gradually increases to 2,740 fathoms at 500, and 2,950 fathoms at 750 miles from Teneriffe. The bottom in these two soundings might have been called 'grey ooze,' for although its nature has altered entirely from the *Globigerina* ooze, the red clay into which it is rapidly passing still contains a considerable admixture of carbonate of lime.

"The depth goes on increasing to a distance of 1,150 miles from Teneriffe, when it reaches 3,150 fathoms ; there the clay is pure and smooth, and contains scarcely a trace of lime. From this great depth the bottom gradually rises, and, with decreasing depth, the grey colour and the calcareous composition of the ooze return. Three soundings in 2,050, 1,900, and 1,950 fathoms on the 'Dolphin Rise' gave highly characteristic examples of the *Globigerina* formation. Passing from the middle plateau of the Atlantic into the western trough, with depths a little over 3,000 fathoms, the red clay returned in all its purity ; and our last sounding, in 1,420 fathoms, before reaching Sombrero, restored the *Globigerina* ooze with its peculiar associated fauna.

"This section shows also the wide extension and the vast geological importance of the red clay formation. The total distance from Teneriffe to Sombrero is about 2,700 miles. Proceeding from east to west, we have—

About   80 miles of volcanic mud and sand,
,,   350   ,,   *Globigerina* ooze,
,,   1,050   ,,   red clay,
,,   330   ,,   *Globigerina* ooze
,   850   ,,   red clay,
,,   40   ,,   *Globigerina* ooze ;

giving a total of 1,900 miles of red clay to 720 miles of *Globigerina* ooze.

" The nature and origin of this vast deposit of clay is a question of the very greatest interest; and although I think there can be no doubt that it is in the main solved, yet some matters of detail are still involved in difficulty. My first impression was that it might be the most minutely divided material, the ultimate sediment produced by the disintegration of the land, by rivers and by the action of the sea on exposed coasts, and held in suspension and distributed by ocean currents, and only making itself manifest in places unoccupied by the *Globigerina* ooze. Several circumstances seemed, however, to negative this mode of origin. The formation seemed too uniform : wherever we met with it, it had the same character, and it only varied in composition in containing less or more carbonate of lime.

" Again, we were gradually becoming more and more convinced that all the important elements of the *Globigerina* ooze lived on the surface, and it seemed evident that, so long as the condition on the surface remained the same, no alteration of contour at the bottom could possibly prevent its accumulation ; and the surface conditions in the Mid-Atlantic were very uniform, a moderate current of a very equal temperature passing continuously over elevations and depressions, and everywhere yielding to the tow-net the ooze-forming *Foraminifera* in the same proportion. The Mid-Atlantic swarms with pelagic *Mollusca*, and, in moderate depths, the shells of these are constantly mixed with the *Globigerina* ooze, sometimes in number sufficient to make up a considerable portion of its bulk. It is clear that these shells must fall in equal numbers upon the red clay, but scarcely a trace of one of them is ever brought up by the dredge on the red clay area. It might be possible to explain the absence of shell-secreting animals living on the bottom, on the supposition that the nature of the deposit was injurious to them ; but then the idea of a current sufficiently strong to sweep them away is negatived by the extreme fineness of the sediment which is being laid down ; the absence of surface shells appears to be intelligible only on the supposition that they are in some way removed.

" We conclude, therefore, that the 'red clay' is not an additional substance introduced from without, and occupying certain depressed regions on account of some law regulating its deposition, but that it is produced by the removal, by some means or other, over these areas, of the carbonate of lime, which forms probably about 98 per cent. of the material of the *Globigerina* ooze. We can trace, indeed, every successive stage in the removal of the carbonate of lime in descending the slope of the ridge or plateau where the *Globigerina* ooze is forming, to the region of the clay. We find, first, that the shells of pteropods and other surface *Mollusca* which are constantly falling on the bottom, are absent, or, if a few remain, they are brittle and yellow, and evidently decaying rapidly. These shells of *Mollusca* decompose more easily and disappear sooner than the smaller, and apparently more delicate, shells of rhizopods. The smaller *Foraminifera* now give way, and are found in lessening proportion to the larger ; the coccoliths first lose their thin outer border and then disappear ; and the clubs of the rhabdoliths get worn out of shape, and are last seen, under a high power, as infinitely minute cylinders scattered over the field. The larger *Foraminifera* are attacked, and instead of being vividly white and delicately sculptured, they become brown and worn, and finally they break up, each according to its fashion ; the chamber-walls of *Globigerina* fall into wedge-shaped pieces, which quickly disappear, and a thick rough crust breaks away from the surface of *Orbulina*, leaving a thin inner sphere, at first beautifully transparent, but soon becoming opaque and crumbling away.

"In the meantime the proportion of the amorphous ' red clay ' to the calcareous elements of all kinds increases, until the latter disappear, with the exception of a few scattered shells of the larger *Foraminifera*, which are still found even in the most characteristic samples of the ' red clay.'

"There seems to be no room left for doubt that the red clay is essentially the insoluble residue, the *ash*, as it were, of the calcareous organisms which form the *Globigerina* ooze, after the calcareous matter has been by some means removed. An ordinary mixture of calcareous *Foraminifera* with the shells of pteropods, forming a fair sample of *Globigerina* ooze from near St. Thomas, was carefully washed, and subjected by Mr.

Buchanan to the action of weak acid ; and he found that there remained after the carbonate of lime had been removed, about 1 per cent. of a reddish mud, consisting of silica, alumina, and the red oxide of iron. This experiment has been frequently repeated with different samples of *Globigerina* ooze, and always with the result that a small proportion of a red sediment remains, which possesses all the characters of the red clay."

*   *   *   *   *

" It seems evident from the observations here recorded, that *clay*, which we have hitherto looked upon as essentially the product of the disintegration of older rocks, may be, under certain circumstances, an organic formation like chalk ; that, as a matter of fact, an area on the surface of the globe, which we have shown to be of vast extent, although we are still far from having ascertained its limits, is being covered by such a deposit at the present day.

" It is impossible to avoid associating such a formation with the fine, smooth, homogeneous clays and schists, poor in fossils, but showing worm-tubes and tracks, and bunches of doubtful branching things, such as Oldhamia, silicious sponges, and thin-shelled peculiar shrimps. Such formations, more or less metamorphosed, are very familiar, especially to the student of palæozoic geology, and they often attain a vast thickness. One is inclined, from the great resemblance between them in composition and in the general character of the included fauna, to suspect that these may be organic formations, like the modern red clay of the Atlantic and Southern Sea, accumulations of the insoluble ashes of shelled creatures.

" The dredging in the red clay on the 13th of March was unusually rich. The bag contained examples, those with calcareous shells rather stunted, of most of the characteristic deep-water groups of the Southern Sea, including *Umbellularia, Euplectella, Pterocrinus, Brisinga, Ophioglypha, Pourtalesia,* and one or two *Mollusca.* This is, however, very rarely the case. Generally the red clay is barren, or contains only a very small number of forms.

It must be admitted that it is very difficult, at

present, to frame any satisfactory explanation of the mode of origin of this singular deposit of red clay.

I cannot say that the theory put forward tentatively, and with much reservation by Professor Thomson, that the calcareous matter is dissolved out by the relatively fresh water of the deep currents from the Antarctic regions, appears satisfactory to me. Nor do I see my way to the acceptance of the suggestion of Dr. Carpenter, that the red clay is the result of the decomposition of previously-formed greensand. At present there is no evidence that greensand casts are ever formed at great depths; nor has it been proved that *Glauconite* is decomposable by the agency of water and carbonic acid.

I think it probable that we shall have to wait some time for a sufficient explanation of the origin of the abyssal red clay, no less than for that of the sublittoral greensand in the intermediate zone. But the importance of the establishment of the fact that these various deposits are being formed in the ocean, at the present day, remains the same, whether its *rationale* be understood or not.

For, suppose the globe to be evenly covered with sea, to a depth say of a thousand fathoms—then, whatever might be the mineral matter composing the sea-bottom, little or no deposit would be formed upon it, the abrading and denuding action of water, at such a depth, being exceedingly slight.

Next, imagine sponges, *Radiolaria, Foraminifera,*
and diatomaceous plants, such as those which now
exist in the deep-sea, to be introduced : they
would be distributed according to the same laws
as at present, the sponges (and possibly some of
the *Foraminifera*) covering the bottom, while other
Foraminifera, with the *Radiolaria* and *Diatomaceœ,*
would increase and multiply in the surface waters.
In accordance with the existing state of things,
the *Radiolaria* and Diatoms would have a universal
distribution, the latter gathering most thickly in
the polar regions, while the *Foraminifera* would
be largely, if not exclusively, confined to the intermediate zone ; and, as a consequence of this distribution, a bed of " chalk " would begin to form in
the intermediate zone, while caps of silicious rock
would accumulate on the circumpolar regions.

Suppose, further, that a part of the intermediate
area were raised to within two or three hundred
fathoms of the surface—for anything that we know
to the contrary, the change of level might determine the substitution of greensand for the
" chalk " ; while, on the other hand, if part of
the same area were depressed to three thousand
fathoms, that change might determine the substitution of a different silicate of alumina and iron—
namely, clay—for the " chalk " that would otherwise be formed.

If the *Challenger* hypothesis, that the red
clay is the residue left by dissolved *Foraminiferous*

skeletons, is correct, then all these deposits alike would be directly, or indirectly, the product of living organisms. But just as a silicious deposit may be metamorphosed into opal or quartzite, and chalk into marble, so known metamorphic agencies may metamorphose clay into schist, clay-slate, slate, gneiss, or even granite. And thus, by the agency of the lowest and simplest of organisms, our imaginary globe might be covered with strata, of all the chief kinds of rock of which the known crust of the earth is composed, of indefinite thickness and extent.

The bearing of the conclusions which are now either established, or highly probable, respecting the origin of silicious, calcareous, and clayey rocks, and their metamorphic derivatives, upon the archæology of the earth, the elucidation of which is the ultimate object of the geologist, is of no small importance.

A hundred years ago the singular insight of Linnæus enabled him to say that "fossils are not the children but the parents of rocks," [1] and the

---

[1] "Petrificata montium calcariorum non filii sed parentes sunt, cum omnis calx oriatur ab animalibus."—*Systema Naturæ*, Ed. xii., t. iii., p. 154. It must be recollected that Linnæus included silex, as well as limestone, under the name of "calx," and that he would probably have arranged Diatoms among animals, as part of "chaos." Ehrenberg quotes another even more pithy passage, which I have not been able to find in any edition of the *Systema* accessible to me : "Sic lapides ab animalibus, nec vice versa. Sic rupes saxei non primævi, sed temporis filiæ."

whole effect of the discoveries made since his time has been to compile a larger and larger commentary upon this text. It is, at present, a perfectly tenable hypothesis that all silicious and calcareous rocks are either directly, or indirectly, derived from material which has, at one time or other, formed part of the organized framework of living organisms. Whether the same generalization may be extended to aluminous rocks, depends upon the conclusion to be drawn from the facts respecting the red clay areas brought to light by the *Challenger*. If we accept the view taken by Wyville Thomson and his colleagues—that the red clay is the residuum left after the calcareous matter of the *Globigerinæ* ooze has been dissolved away—then clay is as much a product of life as limestone, and all known derivatives of clay may have formed part of animal bodies.

So long as the *Globigerinæ*, actually collected at the surface, have not been demonstrated to contain the elements of clay, the *Challenger* hypothesis, as I may term it, must be accepted with reserve and provisionally, but, at present, I cannot but think that it is more probable than any other suggestion which has been made.

Accepting it provisionally, we arrive at the remarkable result that all the chief known constituents of the crust of the earth may have formed part of living bodies; that they may be the "ash" of protoplasm; that the "*rupes saxei*" are not only "*temporis*," but "*vitæ filiæ*"; and, consequently, that the time during which life has been active on the globe may be indefinitely greater than the period, the commencement of which is marked by the oldest known rocks, whether fossiliferous or unfossiliferous.

And thus we are led to see where the solution of a great problem and apparent paradox of geology may lie. Satisfactory evidence now exists that some animals in the existing world have been derived by a process of gradual modification from pre-existing forms. It is undeniable, for example, that the evidence in favour of the derivation of the horse from the later tertiary *Hipparion*, and that of the *Hipparion* from *Anchitherium*, is as complete and cogent as such evidence can reasonably be expected to be; and the further investigations into the history of the tertiary mammalia are pushed, the greater is the accumulation of evidence having the same tendency. So far from palæontology lending no support to the doctrine of evolution—as one sees constantly asserted—that doctrine, if it had no other support, would have been irresistibly forced upon us by the palæontological discoveries of the last twenty years.

If, however, the diverse forms of life which now exist have been produced by the modification of previously-existing less divergent forms, the recent and extinct species, taken as a whole, must fall into series which must converge as we go back in

time. Hence, if the period represented by the
rocks is greater than, or co-extensive with, that
during which life has existed, we ought, somewhere among the ancient formations, to arrive at
the point to which all these series converge, or
from which, in other words, they have diverged—
the primitive undifferentiated protoplasmic living
things, whence the two great series of plants and
animals have taken their departure.

But, as a matter of fact, the amount of convergence of series, in relation to the time occupied by
the deposition of geological formations, is extraordinarily small. Of all animals the higher
Vertebrata are the most complex; and among
these the carnivores and hoofed animals (*Ungulata*)
are highly differentiated. Nevertheless, although
the different lines of modification of the *Carnivora*
and those of the *Ungulata*, respectively, approach
one another, and, although each group is represented by less differentiated forms in the older
tertiary rocks than at the present day, the oldest
tertiary rocks do not bring us near the primitive
form of either. If, in the same way, the convergence of the varied forms of reptiles is measured
against the time during which their remains are
preserved—which is represented by the whole of
the tertiary and mesozoic formations—the amount
of that convergence is far smaller than that of the
lines of mammals, between the present time and
the beginning of the tertiary epoch. And it is a

broad fact that, the lower we go in the scale of organization, the fewer signs are there of convergence towards the primitive form from whence all must have diverged, if evolution be a fact. Nevertheless, that it is a fact in some cases, is proved, and I, for one, have not the courage to suppose that the mode in which some species have taken their origin is different from that in which the rest have originated.

What, then, has become of all the marine animals which, on the hypothesis of evolution, must have existed in myriads in those seas, wherein the many thousand feet of Cambrian and Laurentian rocks now devoid, or almost devoid, of any trace of life were deposited ?

Sir Charles Lyell long ago suggested that the azoic character of these ancient formations might be due to the fact that they had undergone extensive metamorphosis; and readers of the " Principles of Geology " will be familiar with the ingenious manner in which he contrasts the theory of the Gnome, who is acquainted only with the interior of the earth, with those of ordinary philosophers, who know only its exterior.

The metamorphism contemplated by the great modern champion of rational geology is, mainly, that brought about by the exposure of rocks to subterranean heat ; and where no such heat could be shown to have operated, his opponents assumed that no metamorphosis could have taken place. But the formation of greensand, and still more that of the "red clay" (if the *Challenger* hypothesis be correct) affords an insight into a new kind of metamorphosis—not igneous, but aqueous—by which the primitive nature of a deposit may be masked as completely as it can be by the agency of heat. And, as Wyville Thomson suggests, in the passage I have quoted above (p. 17), it further enables us to assign a new cause for the occurrence, so puzzling hitherto, of thousands of feet of unfossiliferous fine-grained schists and slates, in the midst of formations deposited in seas which certainly abounded in life. If the great deposit of "red clay" now forming in the eastern valley of the Atlantic were metamorphosed into slate and then upheaved, it would constitute an "azoic" rock of enormous extent. And yet that rock is now forming in the midst of a sea which swarms with living beings, the great majority of which are provided with calcareous or silicious shells and skeletons; and, therefore, are such as, up to this time, we should have termed eminently preservable.

Thus the discoveries made by the *Challenger* expedition, like all recent advances in our knowledge of the phenomena of biology, or of the changes now being effected in the structure of the surface of the earth, are in accordance with, and lend strong support to, that doctrine of Uniformitarianism, which, fifty

years ago, was held only by a small minority of
English geologists—Lyell, Scrope, and De la Beche
—but now, thanks to the long-continued labours
of the first two, and mainly to those of Sir Charles
Lyell, has gradually passed from the position of a
heresy to that of catholic doctrine.

Applied within the limits of the time registered
by the known fraction of the crust of the earth,
I believe that uniformitarianism is unassailable.
The evidence that, in the enormous lapse of time
between the deposition of the lowest Laurentian
strata and the present day, the forces which have
modified the surface of the crust of the earth were
different in kind, or greater in the intensity of
their action, than those which are now occupied in
the same work, has yet to be produced.   Such
evidence as we possess all tends in the contrary
direction, and is in favour of the same slow and
gradual changes occurring then as now.

But this conclusion in nowise conflicts with the
deductions of the physicist from his no less clear
and certain data.   It may be certain that this
globe has cooled down from a condition in which
life could not have existed ; it may be certain that,
in so cooling, its contracting crust must have
undergone sudden convulsions, which were to our
earthquakes as an earthquake is to the vibration
caused by the periodical eruption of a Geyser ; but
in that case, the earth must, like other respectable
parents, have sowed her wild oats, and got through

her turbulent youth, before we, her children, have any knowledge of her.

So far as the evidence afforded by the superficial crust of the earth goes, the modern geologist can, *ex animo*, repeat the saying of Hutton, " We find no vestige of a beginning—no prospect of an end." However, he will add, with Hutton, " But in thus tracing back the natural operations which have succeeded each other, and mark to us the course of time past, we come to a period in which we cannot see any further." And if he seek to peer into the darkness of this period, he will welcome the light proffered by physics and mathematics.

IV

YEAST

[1871]

IT has been known, from time immemorial, that
the sweet liquids which may be obtained by expressing the juices of the fruits and stems of
various plants, or by steeping malted barley in hot
water, or by mixing honey with water—are liable
to undergo a series of very singular changes, if
freely exposed to the air and left to themselves, in
warm weather. However clear and pellucid the
liquid may have been when first prepared, however
carefully it may have been freed, by straining and
filtration, from even the finest visible impurities,
it will not remain clear. After a time it will
become cloudy and turbid; little bubbles will be
seen rising to the surface, and their abundance will
increase until the liquid hisses as if it were sim-
mering on the fire. By degrees, some of the solid
particles which produce the turbidity of the liquid

collect at its surface into a scum, which is blown
up by the emerging air-bubbles into a thick, foamy
froth. Another moiety sinks to the bottom, and
accumulates as a muddy sediment, or " lees."
When this action has continued, with more or
less violence, for a certain time, it gradually
moderates. The evolution of bubbles slackens,
and finally comes to an end ; scum and lees alike
settle at the bottom, and the fluid is once more
clear and transparent. But it has acquired
properties of which no trace existed in the
original liquid. Instead of being a mere sweet
fluid, mainly composed of sugar and water, the
sugar has more or less completely disappeared ; and
it has acquired that peculiar smell and taste which
we call " spirituous." Instead of being devoid of
any obvious effect upon the animal economy, it
has become possessed of a very wonderful influence
on the nervous system ; so that in small doses it
exhilarates, while in larger it stupefies, and may
even destroy life.

Moreover, if the original fluid is put into a still,
and heated moderately, the first and last product
of its distillation is simple water ; while, when the
altered fluid is subjected to the same process, the
matter which is first condensed in the receiver is
found to be a clear, volatile substance, which is
lighter than water, has a pungent taste and smell,
possesses the intoxicating powers of the fluid in
an eminent degree, and takes fire the moment it

is brought in contact with a flame. The Alchemists called this volatile liquid, which they obtained from wine, "spirits of wine," just as they called hydrochloric acid "spirits of salt," and as we, to this day, call refined turpentine "spirits of turpentine." As the "spiritus," or breath, of a man was thought to be the most refined and subtle part of him, the intelligent essence of man was also conceived as a sort of breath, or spirit; and, by analogy, the most refined essence of anything was called its "spirit." And thus it has come about that we use the same word for the soul of man and for a glass of gin.

At the present day, however, we even more commonly use another name for this peculiar liquid—namely, "alcohol," and its origin is not less singular. The Dutch physician, Van Helmont, lived in the latter part of the sixteenth and the beginning of the seventeenth century—in the transition period between alchemy and chemistry —and was rather more alchemist than chemist. Appended to his "Opera Omnia," published in 1707, there is a very needful "Clavis ad obscuriorum sensum referendum," in which the following passage occurs :—

"ALCOHOL.—Chymicis est liquor aut pulvis summè subtilisatus, vocabulo Orientalibus quoque, cum primis Habessinis, familiari, quibus *cohol* speciatim pulverem impalpabilem ex antimonio pro oculis tingendis denotat . . . Hodie autem, ob analogiam, quivis pulvis tenerior ut pulvis oculorum cancri

summè subtilisatus *alcohol* audit, haud aliter ac spiritus rectifi-
catissimi *alcolisati* dicuntur."

Similarly, Robert Boyle speaks of a fine powder
as "alcohol"; and, so late as the middle of the
last century, the English lexicographer, Nathan
Bailey, defines "alcohol" as "the pure substance
of anything separated from the more gross, a very
fine and impalpable powder, or a very pure, well-
rectified spirit." But, by the time of the publication of Lavoisier's "Traité Elémentaire de
Chimie," in 1789, the term "alcohol," "alkohol,"
or "alkool" (for it is spelt in all three ways), which
Van Helmont had applied primarily to a fine
powder, and only secondarily to spirits of wine, had
lost its primary meaning altogether; and, from
the end of the last century until now, it has, I
believe, been used exclusively as the denotation of
spirits of wine, and bodies chemically allied to that
substance.

The process which gives rise to alcohol in a
saccharine fluid is known to us as "fermentation";
a term based upon the apparent boiling up or
"effervescence" of the fermenting liquid, and of
Latin origin.

Our Teutonic cousins call the same process
"gähren," "gäsen," "göschen," and "gischen";
but, oddly enough, we do not seem to have
retained their verb or their substantive denoting the action itself, though we do use names
identical with, or plainly derived from, theirs for

the scum and lees. These are called, in Low German, "gäscht" and "gischt"; in Anglo-Saxon, "gest," "gist," and "yst," whence our "yeast." Again, in Low German and in Anglo-Saxon there is another name for yeast, having the form "barm," or "beorm"; and, in the Midland Counties, "barm" is the name by which yeast is still best known. In High German, there is a third name for yeast, "hefe," which is not represented in English, so far as I know.

All these words are said by philologers to be derived from roots expressive of the intestine motion of a fermenting substance. Thus "hefe" is derived from "heben," to raise; "barm" from "beren" or "bären," to bear up; "yeast," "yst," and "gist," have all to do with seething and foam, with "yeasty" waves, and "gusty" breezes.

The same reference to the swelling up of the fermenting substance is seen in the Gallo-Latin terms "levure" and "leaven."

It is highly creditable to the ingenuity of our ancestors that the peculiar property of fermented liquids, in virtue of which they "make glad the heart of man," seems to have been known in the remotest periods of which we have any record. All savages take to alcoholic fluids as if they were to the manner born. Our Vedic forefathers intoxicated themselves with the juice of the "soma"; Noah, by a not unnatural reaction against a superfluity of water, appears to have taken the earliest practicable opportunity of qualifying that which he was obliged to drink; and the ghosts of the ancient Egyptians were solaced by pictures of banquets in which the wine-cup passes round, graven on the walls of their tombs. A knowledge of the process of fermentation, therefore, was in all probability possessed by the prehistoric populations of the globe; and it must have become a matter of great interest even to primæval wine-bibbers to study the methods by which fermented liquids could be surely manufactured. No doubt it was soon discovered that the most certain, as well as the most expeditious, way of making a sweet juice ferment was to add to it a little of the scum, or lees, of another fermenting juice. And it can hardly be questioned that this singular excitation of fermentation in one fluid, by a sort of infection, or inoculation, of a little ferment taken from some other fluid, together with the strange swelling, foaming, and hissing of the fermented substance, must have always attracted attention from the more thoughtful. Nevertheless, the commencement of the scientific analysis of the phenomena dates from a period not earlier than the first half of the seventeenth century.

At this time, Van Helmont made a first step, by pointing out that the peculiar hissing and bubbling of a fermented liquid is due, not to the evolution of common air (which he, as the inventor

of the term " gas," calls " gas ventosum "), but to
that of a peculiar kind of air such as is occasionally
met with in caves, mines, and wells, and which
he calls " gas sylvestre."

But a century elapsed before the nature of this
" gas sylvestre," or, as it was afterwards called,
" fixed air," was clearly determined, and it was
found to be identical with that deadly " choke-
damp" by which the lives of those who descend
into old wells, or mines, or brewers' vats, are
sometimes suddenly ended ; and with the poisonous
aëriform fluid which is produced by the combustion of charcoal, and now goes by the name of
carbonic acid gas.

During the same time it gradually became
evident that the presence of sugar was essential to
the production of alcohol and the evolution of
carbonic acid gas, which are the two great and
conspicuous products of fermentation. And finally,
in 1787, the Italian chemist, Fabroni, made the
capital discovery that the yeast ferment, the
presence of which is necessary to fermentation,
is what he termed a " vegeto-animal " substance ;
that is, a body which gives off ammoniacal salts
when it is burned, and is, in other ways, similar
to the gluten of plants and the albumen and
casein of animals.

These discoveries prepared the way for the
illustrious Frenchman, Lavoisier, who first approached the problem of fermentation with a

complete conception of the nature of the work to
be done. The words in which he expresses this
conception, in the treatise on elementary chemistry
to which reference has already been made, mark
the year 1789 as the commencement of a revolution of not less moment in the world of science
than that which simultaneously burst over the
political world, and soon engulfed Lavoisier himself
in one of its mad eddies.

"We may lay it down as an incontestable axiom that, in all
the operations of art and nature, nothing is created ; an equal
quantity of matter exists both before and after the experiment :
the quality and quantity of the elements remain precisely the
same, and nothing takes place beyond changes and modifications
in the combinations of these elements. Upon this principle the
whole art of performing chemical experiments depends ; we
must always suppose an exact equality between the elements of
the body examined and those of the products of its analysis.

"Hence, since from must of grapes we procure alcohol and
carbonic acid, I have an undoubted right to suppose that must
consists of carbonic acid and alcohol. From these premises we
have two modes of ascertaining what passes during vinous fermentation : either by determining the nature of, and the elements
which compose, the fermentable substances ; or by accurately examining the products resulting from fermentation ; and it is evident that the knowledge of either of these must lead to accurate
conclusions concerning the nature and composition of the other.
From these considerations it became necessary accurately to
determine the constituent elements of the fermentable substances ; and for this purpose I did not make use of the compound juices of fruits, the rigorous analysis of which is perhaps
impossible, but made choice of sugar, which is easily analysed,
and the nature of which I have already explained. This substance is a true vegetable oxyd, with two bases, composed of

hydrogen and carbon, brought to the state of an oxyd by means of a certain proportion of oxygen ; and these three elements are combined in such a way that a very slight force is sufficient to destroy the equilibrium of their connection."

After giving the details of his analysis of sugar and of the products of fermentation, Lavoisier continues :—

"The effect of the vinous fermentation upon sugar is thus reduced to the mere separation of its elements into two portions ; one part is oxygenated at the expense of the other, so as to form carbonic acid ; while the other part, being disoxygenated in favour of the latter, is converted into the combustible substance called alkohol ; therefore, if it were possible to re-unite alkohol and carbonic acid together, we ought to form sugar." [1]

Thus Lavoisier thought he had demonstrated that the carbonic acid and the alcohol which are produced by the process of fermentation, are equal in weight to the sugar which disappears; but the application of the more refined methods of modern chemistry to the investigation of the products of fermentation by Pasteur, in 1860, proved that this is not exactly true, and that there is a deficit of from 5 to 7 per cent. of the sugar which is not covered by the alcohol and carbonic acid evolved. The greater part of this deficit is accounted for by the discovery of two substances, glycerine and succinic acid, of the existence of which Lavoisier was unaware, in the

[1] *Elements of Chemistry*. By M. Lavoisier. Translated by Robert Kerr. Second Edition, 1793 (pp. 186—196).

fermented liquid. But about $1\frac{1}{2}$ per cent. still remains to be made good. According to Pasteur, it has been appropriated by the yeast, but the fact that such appropriation takes place cannot be said to be actually proved.

However this may be, there can be no doubt that the constituent elements of fully 98 per cent. of the sugar which has vanished during fermentation have simply undergone rearrangement; like the soldiers of a brigade, who at the word of command divide themselves into the independent regiments to which they belong. The brigade is sugar, the regiments are carbonic acid, succinic acid, alcohol, and glycerine.

From the time of Fabroni, onwards, it has been admitted that the agent by which this surprising rearrangement of the particles of the sugar is effected is the yeast. But the first thoroughly conclusive evidence of the necessity of yeast for the fermentation of sugar was furnished by Appert, whose method of preserving perishable articles of food excited so much attention in France at the beginning of this century. Gay-Lussac, in his " Mémoire sur la Fermentation," [1] alludes to Appert's method of preserving beer-wort unfermented for an indefinite time, by simply boiling the wort and closing the vessel in which the boiling fluid is contained, in such a way as thoroughly to exclude air; and he

[1] *Annales de Chimie*, 1810.

shows that, if a little yeast be introduced into such wort, after it has cooled, the wort at once begins to ferment, even though every precaution be taken to exclude air. And this statement has since received full confirmation from Pasteur.

On the other hand, Schwann, Schroeder and Dusch, and Pasteur, have amply proved that air may be allowed to have free access to beer-wort, without exciting fermentation, if only efficient precautions are taken to prevent the entry of particles of yeast along with the air.

Thus, the truth that the fermentation of a simple solution of sugar in water depends upon the presence of yeast, rests upon an unassailable foundation; and the inquiry into the exact nature of the substance which possesses such a wonderful chemical influence becomes profoundly interesting.

The first step towards the solution of this problem was made two centuries ago by the patient and painstaking Dutch naturalist, Leeuwenhoek, who in the year 1680 wrote thus :—

"Sæpissime examinavi fermentum cerevisiæ, semperque hoc ex globulis per materiam pellucidam fluitantibus, quam cerevisiam esse censui, constare observavi : vidi etiam evidentissime, unumquemque hujus fermenti globulum denuo ex sex distinctis globulis constare, accurate eidem quantitate et formæ, cui globulis sanguinis nostri, respondentibus.

"Verum talis mihi de horum origine et formatione conceptus formabam ; globulis nempe ex quibus farina Tritici, Hordei, Avenæ, Fagotritici, se constat aquæ calore dissolvi et aquæ com-

misceri ; hac, vero aqua, quam cerevisiam vocare licet, refriges-
cente, multos ex minimis particulis in cerevisia coadunari, et hoc
pacto efficere particulam sive globulum, quæ sexta pars est
globuli fæcis, et iterum sex ex hisce globulis conjungi." [1]

Thus Leeuwenhoek discovered that yeast consists of globules floating in a fluid ; but he thought
that they were merely the starchy particles of the
grain from which the wort was made, rearranged.
He discovered the fact that yeast had a definite
structure, but not the meaning of the fact. A
century and a half elapsed, and the investigation
of yeast was recommenced almost simultaneously
by Cagniard de la Tour in France, and by Schwann
and Kützing in Germany. The French observer
was the first to publish his results ; and the subject received at his hands and at those of his
colleague, the botanist Turpin, full and satisfactory
investigation.

The main conclusions at which they arrived are
these. The globular, or oval, corpuscles which
float so thickly in the yeast as to make it muddy,
though the largest are not more than one two-
thousandth of an inch in diameter, and the
smallest may measure less than one seven-
thousandth of an inch, are living organisms. They
multiply with great rapidity by giving off minute
buds, which soon attain the size of their parent,
and then either become detached or remain
united, forming the compound globules of which

[1] Leeuwenhoek, *Arcana Naturæ Detecta.* Ed. Nov., 1721.

Leeuwenhoek speaks, though the constancy of
their arrangement in sixes existed only in the
worthy Dutchman's imagination.

It was very soon made out that these yeast
organisms, to which Turpin gave the name of
Torula cerevisiæ, were more nearly allied to the
lower Fungi than to anything else. Indeed
Turpin, and subsequently Berkeley and Hoffmann,
believed that they had traced the development of
the *Torula* into the well-known and very common
mould—the *Penicillium glaucum*. Other observers
have not succeeded in verifying these statements;
and my own observations lead me to believe, that
while the connection between *Torula* and the
moulds is a very close one, it is of a different
nature from that which has been supposed. I
have never been able to trace the development of
Torula into a true mould ; but it is quite easy to
prove that species of true mould, such as *Penicillium*, when sown in an appropriate nidus, such
as a solution of tartrate of ammonia and yeast-
ash, in water, with or without sugar, give rise to
Torulæ, similar in all respects to *T. cerevisiæ*,
except that they are, on the average, smaller.
Moreover, Bail has observed the development of a
Torula larger than *T. cerevisiæ*, from a *Mucor*, a
mould allied to *Penicillium*.

It follows, therefore, that the *Torulæ*, or
organisms of yeast, are veritable plants ; and conclusive experiments have proved that the power

which causes the rearrangement of the molecules
of the sugar is intimately connected with the life
and growth of the plant. In fact, whatever arrests
the vital activity of the plant also prevents it
from exciting fermentation.

Such being the facts with regard to the nature
of yeast, and the changes which it effects in
sugar, how are they to be accounted for? Before
modern chemistry had come into existence, Stahl,
stumbling, with the stride of genius, upon the conception which lies at the bottom of all modern views
of the process, put forward the notion that the
ferment, being in a state of internal motion, communicated that motion to the sugar, and thus
caused its resolution into new substances. And
Lavoisier, as we have seen, adopts substantially
the same view. But Fabroni, full of the then
novel conception of acids and bases and double
decompositions, propounded the hypothesis that
sugar is an oxide with two bases, and the ferment
a carbonate with two bases ; that the carbon of
the ferment unites with the oxygen of the sugar,
and gives rise to carbonic acid ; while the sugar,
uniting with the nitrogen of the ferment, produces a new substance analogous to opium. This
is decomposed by distillation, and gives rise to
alcohol. Next, in 1803, Thénard propounded a
hypothesis which partakes somewhat of the nature
of both Stahl's and Fabroni's views. " I do not
believe with Lavoisier," he says, " that all the

carbonic acid formed proceeds from the sugar. How, in that case, could we conceive the action of the ferment on it ? I think that the first portions of the acid are due to a combination of the carbon of the ferment with the oxygen of the sugar, and that it is by carrying off a portion of oxygen from the last that the ferment causes the fermentation to commence—the equilibrium between the principles of the sugar being disturbed, they combine afresh to form carbonic acid and alcohol."

The three views here before us may be familiarly exemplified by supposing the sugar to be a cardhouse. According to Stahl, the ferment is somebody who knocks the table, and shakes the cardhouse down; according to Fabroni, the ferment takes out some cards, but puts others in their places; according to Thénard, the ferment simply takes a card out of the bottom story, the result of which is that all the others fall.

As chemistry advanced, facts came to light which put a new face upon Stahl's hypothesis, and gave it a safer foundation than it previously possessed. The general nature of these phenomena may be thus stated :—A body, A, without giving to, or taking from, another body B, any material particles, causes B to decompose into other substances, C, D, E, the sum of the weights of which is equal to the weight of B, which decomposes.

Thus, bitter almonds contain two substances,

amygdalin and synaptase, which can be extracted, in a separate state, from the bitter almonds. The amygdalin thus obtained, if dissolved in water, undergoes no change; but if a little synaptase be added to the solution, the amygdalin splits up into bitter almond oil, prussic acid, and a kind of sugar.

A short time after Cagniard de la Tour discovered the yeast plant, Liebig, struck with the similarity between this and other such processes and the fermentation of sugar, put forward the hypothesis that yeast contains a substance which acts upon sugar, as synaptase acts upon amygdalin. And as the synaptase is certainly neither organized nor alive, but a mere chemical substance, Liebig treated Cagniard de la Tour's discovery with no small contempt, and, from that time to the present, has steadily repudiated the notion that the decomposition of the sugar is, in any sense, the result of the vital activity of the *Torula*. But, though the notion that the *Torula* is a creature which eats sugar and excretes carbonic acid and alcohol, which is not unjustly ridiculed in the most surprising paper that ever made its appearance in a grave scientific journal,[1] may be un-

---

[1] "Das enträthselte Geheimniss der geistigen Gährung (Vorläufige briefliche Mittheilung)" is the title of an anonymous contribution to Wöhler and Liebig's *Annalen der Pharmacie* for 1839, in which a somewhat Rabelaisian imaginary description of the organisation of the "yeast animals" and of the manner in which their functions are performed, is given with a

tenable, the fact that the *Torulæ* are alive, and that yeast does not excite fermentation unless it contains living *Torulæ*, stands fast. Moreover, of late years, the essential participation of living organisms in fermentation other than the alcoholic, has been clearly made out by Pasteur and other chemists.

However, it may be asked, is there any necessary opposition between the so-called "vital" and the strictly physico-chemical views of fermentation? It is quite possible that the living *Torula* may excite fermentation in sugar, because it constantly produces, as an essential part of its vital manifestations, some substance which acts upon the sugar, just as the synaptase acts upon the amygdalin. Or it may be, that, without the formation of any such special substance, the physical condition of the living tissue of the yeast plant is sufficient to effect that small disturbance of the equilibrium of the particles of the sugar, which Lavoisier thought sufficient to effect its decomposition.

Platinum in a very fine state of division—known as platinum black, or *noir de platine*—has

circumstantiality worthy of the author of *Gulliver's Travels*. As a specimen of the writer's humour, his account of what happens when fermentation comes to an end may suffice. "Sobald nämlich die Thiere keinen Zucker mehr vorfinden, so fressen sie sich gegenseitig selbst auf, was durch eine eigene Manipulation geschieht; alles wird verdaut bis auf die Eier, welche unverändert durch den Darmkanal hineingehen ; man hat zuletzt wieder gährungsfähige Hefe, nämlich den Saamen der Thiere, der übrig bleibt."

the very singular property of causing alcohol to
change into acetic acid with great rapidity. The
vinegar plant, which is closely allied to the yeast
plant, has a similar effect upon dilute alcohol,
causing it to absorb the oxygen of the air, and
become converted into vinegar ; and Liebig's
eminent opponent, Pasteur, who has done so much
for the theory and the practice of vinegar-making,
himself suggests that in this case—

"La cause du phénomène physique qui accompagne la vie de
la plante réside dans un état physique propre, analogue à celui
du noir de platine.   Mais il est essentiel de remarquer que cet
état physique de la plante est étroitement lié avec la vie de
cette plante." [1]

Now, if the vinegar plant gives rise to the oxi-
dation of alcohol, on account of its merely physical constitution, it is at any rate possible that
the physical constitution of the yeast plant may
exert a decomposing influence on sugar.

But, without presuming to discuss a question
which leads us into the very arcana of chemistry,
the present state of speculation upon the *modus
operandi* of the yeast plant in producing fermentation is represented, on the one hand, by the
Stahlian doctrine, supported by Liebig, according
to which the atoms of the sugar are shaken into
new combinations, either directly by the *Torulæ*,
or indirectly, by some substance formed by them ;

[1] *Études sur les Mycodermes*, Comptes-Rendus, liv., 1862.

and, on the other hand, by the Thénardian doctrine, supported by Pasteur, according to which the yeast plant assimilates part of the sugar, and, in so doing, disturbs the rest, and determines its resolution into the products of fermentation. Perhaps the two views are not so much opposed as they seem at first sight to be.

But the interest which attaches to the influence of the yeast plants upon the medium in which they live and grow does not arise solely from its bearing upon the theory of fermentation. So long ago as 1838, Turpin compared the *Torulæ* to the ultimate elements of the tissues of animals and plants—" Les organes élémentaires de leurs tissus, comparables aux petits végétaux des levures ordinaires, sont aussi les décompositeurs des substances qui les environnent."

Almost at the same time, and, probably, equally guided by his study of yeast, Schwann was engaged in those remarkable investigations into the form and development of the ultimate structural elements of the tissues of animals, which led him to recognise their fundamental identity with the ultimate structural elements of vegetable organisms.

The yeast plant is a mere sac, or " cell," containing a semi-fluid matter, and Schwann's microscopic analysis resolved all living organisms, in the long run, into an aggregation of such sacs or cells, variously modified ; and tended to show, that all, whatever their ultimate complication, begin their existence in the condition of such simple cells.

In his famous "Mikroskopische Untersuchungen" Schwann speaks of *Torula* as a "cell"; and, in a remarkable note to the passage in which he refers to the yeast plant, Schwann says :—

"I have been unable to avoid mentioning fermentation, because it is the most fully and exactly known operation of cells, and represents, in the simplest fashion, the process which is repeated by every cell of the living body."

In other words, Schwann conceives that every cell of the living body exerts an influence on the matter which surrounds and permeates it, analogous to that which a *Torula* exerts on the saccharine solution by which it is bathed. A wonderfully suggestive thought, opening up views of the nature of the chemical processes of the living body, which have hardly yet received all the development of which they are capable.

Kant defined the special peculiarity of the living body to be that the parts exist for the sake of the whole and the whole for the sake of the parts. But when Turpin and Schwann resolved the living body into an aggregation of quasi-independent cells, each, like a *Torula*, leading its own life and having its own laws of growth and development, the aggregation being dominated and kept working towards a definite end only by a certain harmony among these units, or by the superaddition of a controlling apparatus, such as a nervous system, this conception ceased to be tenable. The cell lives for its own sake, as well as for the sake of the whole organism; and the cells which float in the blood, live at its expense, and profoundly modify it, are almost as much independent organisms as the *Torulæ* which float in beer-wort.

Schwann burdened his enunciation of the "cell theory" with two false suppositions; the one, that the structures he called "nucleus"[1] and "cell-wall" are essential to a cell; the other, that cells are usually formed independently of other cells; but, in 1839, it was a vast and clear gain to arrive at the conception, that the vital functions of all the higher animals and plants are the resultant of the forces inherent in the innumerable minute cells of which they are composed, and that each of them is, itself, an equivalent of one of the lowest and simplest of independent living beings—the *Torula*.

From purely morphological investigations, Turpin and Schwann, as we have seen, arrived at the notion of the fundamental unity of structure of living beings. And, before long, the researches of chemists gradually led up to the conception of the fundamental unity of their composition.

So far back as 1803, Thénard pointed out, in

---

[1] [Later investigations have thrown an entirely new light upon the structure and the functional importance of the nucleus; and have proved that Schwann did not over-estimate its importance. 1894.]

most distinct terms, the important fact that yeast
contains a nitrogenous "animal" substance; and
that such a substance is contained in all ferments.
Before him, Fabroni and Fourcroy speak of the
" vegeto-animal " matter of yeast. In 1844 Mulder
endeavoured to demonstrate that a peculiar substance, which he called "protein," was essentially
characteristic of living matter.

In 1846, Payen writes :—

"Enfin, une loi sans exception me semble apparaître dans
les faits nombreux que j'ai observés et conduire à envisager sous
un nouveau jour la vie végétale ; si je ne m'abuse, tout ce que
dans les tissus végétaux la vue directe où amplifiée nous permet
de discerner sous la forme de cellules et de vaisseaux, ne représente
autre chose que les enveloppes protectrices, les réservoirs et les
conduits, à l'aide desquels les corps animés qui les secrètent et les
façonnent, se logent, puisent et charrient leurs aliments, déposent
et isolent les matières excrétées."

And again :—

"Afin de compléter aujourd'hui l'énoncé du fait général, je
rappellerai que les corps, doué des fonctions accomplies dans
les tissus des plantes, sont formés des éléments qui constituent,
en proportion peu variable, les organismes animaux ; qu'ainsi
l'on est conduit à reconnaître une immense unité de composition
élémentaire dans tous les corps vivants de la nature." [1]

In the year (1846) in which these remarkable
passages were published, the eminent German
botanist, Von Mohl, invented the word "protoplasm," as a name for one portion of those nitrogenous contents of the cells of living plants, the

[1] "Mém. sur les Développements des Végétaux," &c.—*Mém.
Présentées.* ix. 1846.

close chemical resemblance of which to the essential constituents of living animals is so strongly
indicated by Payen. And through the twenty-
five years that have passed, since the matter of
life was first called protoplasm, a host of investigators, among whom Cohn, Max Schulze, and
Kühne must be named as leaders, have accumulated evidence, morphological, physiological, and
chemical, in favour of that "immense unité de
composition élémentaire dans tous les corps vivants
de la nature," into which Payen had, so early, a
clear insight.

As far back as 1850, Cohn wrote, apparently
without any knowledge of what Payen had said
before him :—

"The protoplasm of the botanist, and the contractile substance and sarcode of the zoologist, must be, if not identical, yet
in a high degree analogous substances. Hence, from this point
of view, the difference between animals and plants consists in
this ; that, in the latter, the contractile substance, as a primordial
utricle, is enclosed within an inert cellulose membrane, which
permits it only to exhibit an internal motion, expressed by the
phenomena of rotation and circulation, while, in the former, it
is not so enclosed. The protoplasm in the form of the primordial
utricle is, as it were, the animal element in the plant, but
which is imprisoned, and only becomes free in the animal ; or,
to strip off the metaphor which obscures simple thought, the
energy of organic vitality which is manifested in movement is
especially exhibited by a nitrogenous contractile substance,
which in plants is limited and fettered by an inert membrane,
in animals not so." [1]

[1] Cohn, "Ueber Protococcus pluvialis," in the *Nova Acta* for
1850.

In 1868, thinking that an untechnical statement of the views current among the leaders of
biological science might be interesting to the
general public, I gave a lecture embodying them in
Edinburgh. Those who have not made the mistake of attempting to approach biology, either by
the high *à priori* road of mere philosophical speculation, or by the mere low *à posteriori* lane offered
by the tube of a microscope, but have taken the
trouble to become acquainted with well-ascertained
facts and with their history, will not need to
be told that in what I had to say " as regards
protoplasm " in my lecture " On the Physical
Basis of Life " (Vol. I. of these Essays, p. 130),
there was nothing new; and, as I hope, nothing that the present state of knowledge does
not justify us in believing to be true. Under these
circumstances, my surprise may be imagined, when
I found, that the mere statement of facts and of
views, long familiar to me as part of the common
scientific property of Continental workers, raised a
sort of storm in this country, not only by exciting
the wrath of unscientific persons whose pet pre-
judices they seemed to touch, but by giving rise to
quite superfluous explosions on the part of some
who should have been better informed.

Dr. Stirling, for example, made my essay the
subject of a special critical lecture,[1] which I have

[1] Subsequently published under the title of "As regards
Protoplasm."

read with much interest, though, I confess, the meaning of much of it remains as dark to me as does the "Secret of Hegel" after Dr. Stirling's elaborate revelation of it. Dr. Stirling's method of dealing with the subject is peculiar. "Protoplasm" is a question of history, so far as it is a name; of fact, so far as it is a thing. Dr. Stirling has not taken the trouble to refer to the original authorities for his history, which is consequently a travesty; and still less has he concerned himself with looking at the facts, but contents himself with taking them also at second-hand. A most amusing example of this fashion of dealing with scientific statements is furnished by Dr. Stirling's remarks upon my account of the protoplasm of the nettle hair. That account was drawn up from careful and often-repeated observation of the facts. Dr. Stirling thinks he is offering a valid criticism, when he says that my valued friend Professor Stricker gives a somewhat different statement about protoplasm. But why in the world did not this distinguished Hegelian look at a nettle hair for himself, before venturing to speak about the matter at all? Why trouble himself about what either Stricker or I say, when any tyro can see the facts for himself, if he is provided with those not rare articles, a nettle and a microscope? But I suppose this would have been "*Aufklärung*"—a recurrence to the base common-sense philosophy of the eighteenth century, which liked to see before it believed, and to understand before it criticised Dr. Stirling winds up his paper with the following paragraph :—

"In short, the whole position of Mr. Huxley, (1) that all organisms consist alike of the same life-matter, (2) which life-matter is, for its part, due only to chemistry, must be pronounced untenable—nor less untenable (3) the materialism he would found on it."

The paragraph contains three distinct assertions concerning my views, and just the same number of utter misrepresentations of them. That which I have numbered (1) turns on the ambiguity of the word "same," for a discussion of which I would refer Dr. Stirling to a great hero of *"Aufklarung,"* Archbishop Whately ; statement number (2) is, in my judgment, absurd, and certainly I have never said anything resembling it ; while, as to number (3), one great object of my essay was to show that what is called "materialism" has no sound philosophical basis !

As we have seen, the study of yeast has led investigators face to face with problems of immense interest in pure chemistry, and in animal and vegetable morphology. Its physiology is not less rich in subjects for inquiry. Take, for example, the singular fact that yeast will increase indefinitely when grown in the dark, in water containing only tartrate of ammonia, a small percentage of mineral salts, and sugar. Out of these materials the *Torulæ* will manufacture nitrogenous protoplasm, cellulose, and fatty matters, in any quantity, although they are wholly deprived of those rays of the sun, the influence of which is essential to the growth of ordinary plants. There has been a great deal of speculation lately, as to how the living organisms buried beneath two or three thousand fathoms of water, and therefore in all probability almost deprived of light, live. If any of them possess the same powers as yeast (and the same capacity for living without light is exhibited by some other fungi) there would seem to be no difficulty about the matter.

Of the pathological bearings of the study of yeast, and other such organisms, I have spoken elsewhere. It is certain that, in some animals, devastating epidemics are caused by fungi of low order—similar to those of which *Torula* is a sort of offshoot. It is certain that such diseases are propagated by contagion and infection, in just the same way as ordinary contagious and infectious diseases are propagated. Of course, it does not follow from this, that all contagious and infectious diseases are caused by organisms of as definite and independent a character as the *Torula* ; but, I think, it does follow that it is prudent and wise to satisfy one's self in each particular case, that the " germ theory " cannot and will not explain the facts, before having recourse to hypotheses which have no equal support from analogy.

V

ON THE FORMATION OF COAL

[1870]

THE lumps of coal in a coal-scuttle very often have a roughly cubical form. If one of them be picked out and examined with a little care, it will be found that its six sides are not exactly alike. Two opposite sides are comparatively smooth and shining, while the other four are much rougher, and are marked by lines which run parallel with the smooth sides. The coal readily splits along these lines, and the split surfaces thus formed are parallel with the smooth faces. In other words, there is a sort of rough and incomplete stratification in the lump of coal, as if it were a book, the leaves of which had stuck together very closely.

Sometimes the faces along which the coal splits are not smooth, but exhibit a thin layer of dull, charred-looking substance, which is known as "mineral charcoal."

Occasionally one of the faces of a lump of coal will present impressions, which are obviously those of the stem, or leaves, of a plant; but though hard mineral masses of pyrites, and even fine mud, may occur here and there, neither sand nor pebbles are met with.

When the coal burns, the chief ultimate products of its combustion are carbonic acid, water, and ammoniacal products, which escape up the chimney; and a greater or less amount of residual earthy salts, which take the form of ash. These products are, to a great extent, such as would result from the burning of so much wood.

These properties of coal may be made out without any very refined appliances, but the microscope reveals something more. Black and opaque as ordinary coal is, slices of it become transparent if they are cemented in Canada balsam, and rubbed down very thin, in the ordinary way of making thin sections of non-transparent bodies. But as the thin slices, made in this way, are very apt to crack and break into fragments, it is better to employ marine glue as the cementing material. By the use of this substance, slices of considerable size and of extreme thinness and transparency may be obtained.[1]

[1] My assistant in the Museum of Practical Geology, Mr. Newton, invented this excellent method of obtaining thin slices of coal.

Now let us suppose two such slices to be prepared from our lump of coal—one parallel with the bedding, the other perpendicular to it; and let us call the one the horizontal, and the other the vertical, section. The horizontal section will present more or less rounded yellow patches and streaks, scattered irregularly through the dark brown, or blackish, ground substance; while the vertical section will exhibit mere elongated bars and granules of the same yellow materials, disposed in lines which correspond, roughly, with the general direction of the bedding of the coal.

This is the microscopic structure of an ordinary piece of coal. But if a great series of coals, from different localities and seams, or even from different parts of the same seam, be examined, this structure will be found to vary in two directions. In the anthracitic, or stone-coals, which burn like coke, the yellow matter diminishes, and the ground substance becomes more predominant, blacker, and more opaque, until it becomes impossible to grind a section thin enough to be translucent; while, on the other hand, in such as the " Better-Bed " coal of the neighbourhood of Bradford, which burns with much flame, the coal is of a far lighter colour, and transparent sections are very easily obtained. In the browner parts of this coal, sharp eyes will readily detect multitudes of curious little coin-shaped bodies, of a yellowish brown colour, embedded in the dark brown ground substance. On the average, these little brown bodies may have a diameter of about one-twentieth of an inch. They lie with their flat surfaces nearly parallel with the two smooth faces of the block in which they are contained; and, on one side of each, there may be discerned a figure, consisting of three straight linear marks, which radiate from the centre of the disk, but do not quite reach its circumference. In the horizontal section these disks are often converted into more or less complete rings; while in the vertical sections they appear like thick hoops, the sides of which have been pressed together. The disks are, therefore, flattened bags; and favourable sections show that the three-rayed marking is the expression of three clefts, which penetrate one wall of the bag.

The sides of the bags are sometimes closely approximated; but, when the bags are less flattened, their cavities are, usually, filled with numerous, irregularly rounded, hollow bodies, having the same kind of wall as the large ones, but not more than one seven-hundredth of an inch in diameter.

In favourable specimens, again, almost the whole ground substance appears to be made up of similar bodies—more or less carbonized or blackened—and, in these, there can be no doubt that, with the exception of patches of mineral charcoal, here and there, the whole mass of the

coal is made up of an accumulation of the larger
and of the smaller sacs.

But, in one and the same slice, every transition
can be observed from this structure to that which
has been described as characteristic of ordinary
coal. The latter appears to rise out of the
former, by the breaking-up and increasing car-
bonization of the larger and the smaller sacs.
And, in the anthracitic coals, this process appears
to have gone to such a length, as to destroy the
original structure altogether, and to replace it by
a completely carbonized substance.

Thus coal may be said, speaking broadly, to be
composed of two constituents: firstly, mineral
charcoal; and, secondly, coal proper. The nature
of the mineral charcoal has long since been
determined. Its structure shows it to consist of
the remains of the stems and leaves of plants,
reduced a little more than their carbon. Again,
some of the coal is made up of the crushed and
flattened bark, or outer coat, of the stems of plants,
the inner wood of which has completely decayed
away. But what I may term the "saccular
matter" of the coal, which, either in its primary
or in its degraded form, constitutes by far the
greater part of all the bituminous coals I have
examined, is certainly not mineral charcoal; nor
is its structure that of any stem or leaf. Hence
its real nature is, at first, by no means apparent,
and has been the subject of much discussion.

The first person who threw any light upon the problem, as far as I have been able to discover, was the well-known geologist, Professor Morris. It is now thirty-four years since he carefully described and figured the coin-shaped bodies, or larger sacs, as I have called them, in a note appended to the famous paper "On the Coalbrookdale Coal-Field," published at that time, by the present President of the Geological Society, Mr. Prestwich. With much sagacity, Professor Morris divined the real nature of these bodies, and boldly affirmed them to be the spore-cases of a plant allied to the living club-mosses.

But discovery sometimes makes a long halt; and it is only a few years since Mr. Carruthers determined the plant (or rather one of the plants) which produces these spore-cases, by finding the discoidal sacs still adherent to the leaves of the fossilized cone which produced them. He gave the name of *Flemingites gracilis* to the plant of which the cones form a part. The branches and stem of this plant are not yet certainly known, but there is no sort of doubt that it was closely allied to the *Lepidodendron*, the remains of which abound in the coal formation. The *Lepidodendra* were shrubs and trees which put one more in mind of an *Araucaria* than of any other familiar plant; and the ends of the fruiting branches were terminated by cones, or catkins, somewhat like the bodies so named in a fir, or a willow. These conical fruits, however, did not produce seeds; but the leaves of which they were composed bore upon their surfaces sacs full of spores or sporangia, such as those one sees on the under surface of a bracken leaf. Now, it is these sporangia of the Lepidodendroid plant *Flemingites* which were identified by Mr. Carruthers with the free sporangia described by Professor Morris, which are the same as the large sacs of which I have spoken. And, more than this, there is no doubt that the small sacs are the spores, which were originally contained in the sporangia.

The living club-mosses are, for the most part, insignificant and creeping herbs, which, super-ficially, very closely resemble true mosses, and none of them reach more than two or three feet in height. But, in their essential structure, they very closely resemble the earliest Lepidodendroid trees of the coal : their stems and leaves are similar; so are their cones; and no less like are the sporangia and spores; while even in their size, the spores of the *Lepidodendron* and those of the existing *Lycopodium*, or club-moss, very closely approach one another.

Thus, the singular conclusion is forced upon us, that the greater and the smaller sacs of the "Better-Bed" and other coals, in which the primitive structure is well preserved, are simply the sporangia and spores of certain plants, many of which were closely allied to the existing club-mosses. And if, as I believe, it can be demon-strated that ordinary coal is nothing but " saccular" coal which has undergone a certain amount of that alteration which, if continued, would convert it into anthracite; then, the con-clusion is obvious, that the great mass of the coal we burn is the result of the accumulation of the spores and spore-cases of plants, other parts of which have furnished the carbonized stems and the mineral charcoal, or have left their impressions on the surfaces of the layer.

Of the multitudinous speculations which, at various times, have been entertained respecting the origin and mode of formation of coal, several appear to be negatived, and put out of court, by the structural facts the significance of which I have endeavoured to explain. These facts, for example, do not permit us to suppose that coal is an accumulation of peaty matter, as some have held.

Again, the late Professor Quekett was one of the first observers who gave a correct description of what I have termed the " saccular" structure of coal; and, rightly perceiving that this structure was something quite different from that of any known plant, he imagined that it proceeded from some extinct vegetable organism which was peculiarly abundant amongst the coal-forming plants. But this explanation is at once shown to be untenable when the smaller and the larger sacs are proved to be spores or sporangia.

Some, once more, have imagined that coal was of submarine origin; and though the notion is amply and easily refuted by other considerations, it may be worth while to remark, that it is impossible to comprehend how a mass of light and resinous spores should have reached the bottom of the sea, or should have stopped in that position if they had got there.

At the same time, it is proper to remark that I do not presume to suggest that all coal must needs have the same structure; or that there may not be coals in which the proportions of wood and spores, or spore-cases, are very different from those which I have examined. All I repeat is, that none of the coals which have come under my notice have enabled me to observe such a difference. But, according to Principal Dawson, who has so sedulously examined the fossil remains of plants in North America, it is otherwise with the vast accumulations of coal in that country.

"The true coal," says Dr. Dawson, "consists principally of the flattened bark of Sigillarioid and other trees, intermixed with leaves of Ferns and *Cordaites*, and other herbaceous *débris*, and with fragments of decayed wood, constituting 'mineral charcoal,' all these materials having manifestly alike grown and accumulated where we find them." [1]

---

[1] *Acadian Geology*, 2nd edition, p. 138.

When I had the pleasure of seeing Principal
Dawson in London last summer, I showed him
my sections of coal, and begged him to re-examine
some of the American coals on his return to
Canada, with an eye to the presence of spores and
sporangia, such as I was able to show him in our
English and Scotch coals.   He has been good
enough to do so ; and in a letter dated September
26th, 1870, he informs me that—

"Indications of spore-cases are rare, except in certain coarse
shaly coals and portions of coals, and in the roofs of the seams.
The most marked case I have yet met with is the shaly coal
referred to as containing *Sporangites* in my paper on the conditions of accumulation of coal ("Journal of the Geological
Society," vol. xxii. pp. 115, 139, and 165).   The purer coals certainly consist principally of cubical tissues with some true woody
matter, and the spore-cases, &c., are chiefly in the coarse and
shaly layers.   This is my old doctrine in my two papers in the
"Journal of the Geological Society," and I see nothing to modify
it.   Your observations, however, make it probable that the
frequent *clear spots* in the cannels are spore-cases."

Dr. Dawson's results are the more remarkable,
as the numerous specimens of British coal, from
various localities, which I have examined, tell one
tale as to the predominance of the spore and
sporangium element in their composition; and as
it is exactly in the finest and purest coals, such as
the "Better-Bed" coal of Lowmoor, that the
spores and sporangia obviously constitute almost
the entire mass of the deposit.

Coal, such as that which has been described, is

always found in sheets, or "seams," varying from
a fraction of an inch to many feet in thickness,
enclosed in the substance of the earth at very
various depths, between beds of rock of different
kinds. As a rule, every seam of coal rests upon a
thicker, or thinner, bed of clay, which is known
as "under-clay." These alternations of beds of
coal, clay, and rock may be repeated many times,
and are known as the "coal-measures"; and in
some regions, as in South Wales and in Nova
Scotia, the coal-measures attain a thickness of
twelve or fourteen thousand feet, and enclose
eighty or a hundred seams of coal, each with its
under-clay, and separated from those above and
below by beds of sandstone and shale.

The position of the beds which constitute the
coal-measures is infinitely diverse. Sometimes
they are tilted up vertically, sometimes they are
horizontal, sometimes curved into great basins;
sometimes they come to the surface, sometimes
they are covered up by thousands of feet of rock.
But, whatever their present position, there is
abundant and conclusive evidence that every
under-clay was once a surface soil. Not only do
carbonized root-fibres frequently abound in these
under-clays; but the stools of trees, the trunks of
which are broken off and confounded with the bed
of coal, have been repeatedly found passing into
radiating roots, still embedded in the under-clay.
On many parts of the coast of England, what are

commonly known as "submarine forests" are to be seen at low water. They consist, for the most part, of short stools of oak, beech, and fir-trees, still fixed by their long roots in the bed of blue clay in which they originally grew. If one of these submarine forest beds should be gradually depressed and covered up by new deposits, it would present just the same characters as an under-clay of the coal, if the *Sigillaria* and *Lepidodendron* of the ancient world were substituted for the oak, or the beech, of our own times.

In a tropical forest, at the present day, the trunks of fallen trees, and the stools of such trees as may have been broken by the violence of storms, remain entire for but a short time. Contrary to what might be expected, the dense wood of the tree decays, and suffers from the ravages of insects, more swiftly than the bark. And the traveller, setting his foot on a prostrate trunk, finds that it is a mere shell, which breaks under his weight, and lands his foot amidst the insects, or the reptiles, which have sought food or refuge within.

The trees of the coal forests present parallel conditions. When the fallen trunks which have entered into the composition of the bed of coal are identifiable, they are mere double shells of bark, flattened together in consequence of the destruction of the woody core; and Sir Charles

Lyell and Principal Dawson discovered, in the hollow stools of coal trees of Nova Scotia, the remains of snails, millipedes, and salamander-like creatures, embedded in a deposit of a different character from that which surrounded the exterior of the trees. Thus, in endeavouring to comprehend the formation of a seam of coal, we must try to picture to ourselves a thick forest, formed for the most part of trees like gigantic club-mosses, mares'-tails, and tree-ferns, with here and there some that had more resemblance to our existing yews and fir-trees. We must suppose that, as the seasons rolled by, the plants grew and developed their spores and seeds; that they shed these in enormous quantities, which accumulated on the ground beneath; and that, every now and then, they added a dead frond or leaf; or, at longer intervals, a rotten branch, or a dead trunk, to the mass.

A certain proportion of the spores and seeds no doubt fulfilled their obvious function, and, carried by the wind to unoccupied regions, extended the limits of the forest; many might be washed away by rain into streams, and be lost; but a large portion must have remained, to accumulate like beech-mast, or acorns, beneath the trees of a modern forest.

But, in this case, it may be asked, why does not our English coal consist of stems and leaves to a much greater extent than it does? What is the reason of the predominance of the spores and spore-cases in it ?

A ready answer to this question is afforded by the study of a living full-grown club-moss. Shake it upon a piece of paper, and it emits a cloud of fine dust, which falls over the paper, and is the well-known Lycopodium powder. Now this powder used to be, and I believe still is, employed for two objects which seem, at first sight, to have no particular connection with one another. It is, or was, employed in making lightning, and in making pills. The coats of the spores contain so much resinous matter, that a pinch of Lycopodium powder, thrown through the flame of a candle, burns with an instantaneous flash, which has long done duty for lightning on the stage. And the same character makes it a capital coating for pills; for the resinous powder prevents the drug from being wetted by the saliva, and thus bars the nauseous flavour from the sensitive papillæ of the tongue.

But this resinous matter, which lies in the walls of the spores and sporangia, is a substance not easily altered by air and water, and hence tends to preserve these bodies, just as the bituminized cerecloth preserves an Egyptian mummy; while, on the other hand, the merely woody stem and leaves tend to rot, as fast as the wood of the mummy's coffin has rotted. Thus the mixed heap of spores, leaves, and stems in the coal-forest would be persistently searched by the long-continued action of air and rain; the leaves and stems would gradually be reduced to little but their carbon, or, in other words, to the condition of mineral charcoal in which we find them; while the spores and sporangia remained as a comparatively unaltered and compact residuum.

There is, indeed, tolerably clear evidence that the coal must, under some circumstances, have been converted into a substance hard enough to be rolled into pebbles, while it yet lay at the surface of the earth; for in some seams of coal, the courses of rivulets, which must have been living water, while the stratum in which their remains are found was still at the surface, have been observed to contain rolled pebbles of the very coal through which the stream has cut its way.

The structural facts are such as to leave no alternative but to adopt the view of the origin of such coal as I have described, which has just been stated; but, happily, the process is not without analogy at the present day. I possess a specimen of what is called "white coal" from Australia. It is an inflammable material, burning with a bright flame, and having much the consistence and appearance of oat-cake, which, I am informed, covers a considerable area. It consists, almost entirely, of a compacted mass of spores and spore-cases. But the fine particles of blown sand which are scattered through it, show that it must have accumulated, subaërially, upon the surface of a soil covered by a forest of cryptogamous plants, probably tree-ferns.

As regards this important point of the subaerial region of coal, I am glad to find myself in entire accordance with Principal Dawson, who bases his conclusions upon other, but no less forcible, considerations. In a passage, which is the continuation of that already cited, he writes :—

"(3) The microscopical structure and chemical composition of the beds of cannel coal and earthy bitumen, and of the more highly bituminous and carbonaceous shale, show them to have been of the nature of the fine vegetable mud which accumulates in the ponds and shallow lakes of modern swamps. When such fine vegetable sediment is mixed, as is often the case, with clay, it becomes similar to the bituminous limestone and calcareo-bituminous shales of the coal-measures. (4) A few of the under-clays, which support beds of coal, are of the nature of the vegetable mud above referred to ; but the greater part are argillo-arenaceous in composition, with little vegetable matter, and bleached by the drainage from them of water containing the products of vegetable decay. They are, in short, loamy or clay soils, and must have been sufficiently above water to admit of drainage. The absence of sulphurets, and the occurrence of carbonate of iron in connection with them, prove that, when they existed as soils, rain-water, and not sea-water, percolated them. (5) The coal and the fossil forests present many evidences of subaerial conditions. Most of the erect and prostrate trees had become hollow shells of bark before they were finally embedded, and their wood had broken into cubical pieces of mineral charcoal. Land-snails and galley-worms (*Xylobius*) crept into them, and they became dens, or traps, for reptiles. Large quantities of mineral charcoal occur on the surface of all the large beds of coal. None of these appearances could have been produced by subaqueous action. (6) Though the roots of the *Sigillaria* bear more resemblance to the rhizomes of certain aquatic plants; yet, structurally, they are absolutely identical with the roots of Cycads, which the stems also resemble. Further, the *Sigillariæ* grew on the same soils which supported Conifers, *Lepidodendra*, *Cordaites*, and Ferns—plants which could not have grown in water. Again, with the exception perhaps of some *Pinnulariæ* and *Asterophyllites*, there is a remarkable absence from the coal measures of any form of properly aquatic vegetation. (7) The occurrence of marine, or brackish-water animals, in the roofs of coal beds, or even in the coal itself, affords no evidence of subaqueous accumulation, since the same thing occurs in the case of modern submarine forests. For these and other reasons, some of which are more fully stated in the papers already referred to, while I admit that the areas of coal accumulation were frequently submerged, I must maintain that the true coal is a subaerial accumulation by vegetable growth on soils, wet and swampy it is true, but not submerged."

I am almost disposed to doubt whether it is necessary to make the concession of "wet and swampy"; otherwise, there is nothing that I know of to be said against this excellent conspectus of the reasons for believing in the subaerial origin of coal.

But the coal accumulated upon the area covered by one of the great forests of the carboniferous epoch would, in course of time, have been wasted away by the small, but constant, wear and tear of rain and streams, had the land which supported it remained at the same level, or been gradually raised to a greater elevation. And, no doubt, as much coal as now exists has been destroyed, after its formation, in this way. What are now known

as coal districts owe their importance to the fact
that they were areas of slow depression, during a
greater or less portion of the carboniferous epoch;
and that, in virtue of this circumstance, Mother
Earth was enabled to cover up her vegetable
treasures, and preserve them from destruction.

Wherever a coal-field now exists, there must
formerly have been free access for a great river, or
for a shallow sea, bearing sediment in the shape of
sand and mud. When the coal-forest area became
slowly depressed, the waters must have spread
over it, and have deposited their burden upon the
surface of the bed of coal, in the form of layers,
which are now converted into shale, or sandstone.
Then followed a period of rest, in which the
superincumbent shallow waters became completely
filled up, and finally replaced, by fine mud, which
settled down into a new under-clay, and furnished
the soil for a fresh forest growth. This flourished,
and heaped up its spores and wood into coal, until
the stage of slow depression recommenced. And,
in some localities, as I have mentioned, the process
was repeated until the first of the alternating
beds had sunk to near three miles below its
original level at the surface of the earth.

In reflecting on the statement, thus briefly
made, of the main facts connected with the
origin of the coal formed during the carboniferous
epoch, two or three considerations suggest themselves.

In the first place, the great phantom of geological time rises before the student of this, as of all other, fragments of the history of our earth—springing irrepressibly out of the facts, like the Djin from the jar which the fishermen so incautiously opened ; and like the Djin again, being vaporous, shifting, and indefinable, but unmistakably gigantic. However modest the bases of one's calculation may be, the minimum of time assignable to the coal period remains something stupendous.

Principal Dawson is the last person likely to be guilty of exaggeration in this matter, and it will be well to consider what he has to say about it :—

"The rate of accumulation of coal was very slow. The climate of the period, in the northern temperate zone, was of such a character that the true conifers show rings of growth, not larger, nor much less distinct, than those of many of their modern congeners. The *Sigillariæ* and *Calamites* were not, as often supposed, composed wholly, or even principally, of lax and soft tissues, or necessarily short-lived. The former had, it is true, a very thick inner bark ; but their dense woody axis, their thick and nearly imperishable outer bark, and their scanty and rigid foliage, would indicate no very rapid growth or decay. In the case of the *Sigillariæ*, the variations in the leaf-scars in different parts of the trunk, the intercalation of new ridges at the surface representing that of new woody wedges in the axis, the transverse marks left by the stages of upward growth, all indicate that several years must have been required for the growth of stems of moderate size. The enormous roots of these trees, and the condition of the coal-swamps, must have exempted them from the danger of being overthrown by violence. They probably fell in successive generations from natural decay ; and making every allowance for other materials, we may safely assert that every foot of thickness of pure bituminous coal implies the quiet growth and fall of at least fifty generations of *Sigillariæ*, and therefore an undisturbed condition of forest growth enduring through many centuries. Further, there is evidence that an immense amount of loose parenchymatous tissue, and even of wood, perished by decay, and we do not know to what extent even the most durable tissues may have disappeared in this way ; so that, in many coal-seams, we may have only a very small part of the vegetable matter produced."

Undoubtedly the force of these reflections is not diminished when the bituminous coal, as in Britain, consists of accumulated spores and spore-cases, rather than of stems. But, suppose we adopt Principal Dawson's assumption, that one foot of coal represents fifty generations of coal plants; and, further, make the moderate supposition that each generation of coal plants took ten years to come to maturity—then, each foot-thickness of coal represents five hundred years. The super-imposed beds of coal in one coal-field may amount to a thickness of fifty or sixty feet, and therefore the coal alone, in that field, represents $500 \times 50 = 25,000$ years. But the actual coal is but an insignificant portion of the total deposit, which, as has been seen, may amount to between two and three miles of vertical thickness. Suppose it be 12,000 feet—which is 240 times the thickness of the actual coal—is there any reason why we should believe it may not have taken 240 times as long to form ? I know of none. But, in this case, the

time which the coal-field represents would be
25,000 × 240 = 6,000,000 years. As affording a
definite chronology, of course such calculations as
these are of no value ; but they have much use in
fixing one's attention upon a possible minimum.
A man may be puzzled if he is asked how long
Rome took a-building ; but he is proverbially safe
if he affirms it not to have been built in a day ;
and our geological calculations are all, at present,
pretty much on that footing.

A second consideration which the study of the
coal brings prominently before the mind of any one
who is familiar with palæontology is, that the
coal Flora, viewed in relation to the enormous
period of time which it lasted, and to the still
vaster period which has elapsed since it flourished,
underwent little change while it endured, and in
its peculiar characters, differs strangely little from
that which at present exist.

The same species of plants are to be met with
throughout the whole thickness of a coal-field, and
the youngest are not sensibly different from the
oldest. But more than this. Notwithstanding
that the carboniferous period is separated from us
by more than the whole time represented by the
secondary and tertiary formations, the great types
of vegetation were as distinct then as now. The
structure of the modern club-moss furnishes a
complete explanation of the fossil remains of the
Lepidodendra, and the fronds of some of the ancient

ferns are hard to distinguish from existing ones.
At the same time, it must be remembered, that
there is nowhere in the world, at present, any
forest which bears more than a rough analogy with
a coal-forest. The types may remain, but the
details of their form, their relative proportions,
their associates, are all altered. And the tree-fern
forest of Tasmania, or New Zealand, gives one only
a faint and remote image of the vegetation of
the ancient world.

Once more, an invariably-recurring lesson of
geological history, at whatever point its study is
taken up : the lesson of the almost infinite slow-
ness of the modification of living forms. The
lines of the pedigrees of living things break off
almost before they begin to converge.

Finally, yet another curious consideration. Let
us suppose that one of the stupid, salamander-like
Labyrinthodonts, which pottered, with much belly
and little leg, like Falstaff in his old age, among
the coal-forests, could have had thinking power
enough in his small brain to reflect upon the
showers of spores which kept on falling through
years and centuries, while perhaps not one in ten
million fulfilled its apparent purpose, and repro-
duced the organism which gave it birth : surely
he might have been excused for moralizing upon
the thoughtless and wanton extravagance which
Nature displayed in her operations.

But we have the advantage over our shovel-headed predecessor—or possibly ancestor—and can perceive that a certain vein of thrift runs through this apparent prodigality. Nature is never in a hurry, and seems to have had always before her eyes the adage, " Keep a thing long enough, and you will find a use for it." She has kept her beds of coal many millions of years without being able to find much use for them ; she has sent them down beneath the sea, and the sea-beasts could make nothing of them ; she has raised them up into dry land, and laid the black veins bare, and still, for ages and ages, there was no living thing on the face of the earth that could see any sort of value in them; and it was only the other day, so to speak, that she turned a new creature out of her workshop, who by degrees acquired sufficient wits to make a fire, and then to discover that the black rock would burn.

I suppose that nineteen hundred years ago, when Julius Cæsar was good enough to deal with Britain as we have dealt with New Zealand, the primæval Briton, blue with cold and woad, may have known that the strange black stone, of which he found lumps here and there in his wanderings, would burn, and so help to warm his body and cook his food. Saxon, Dane, and Norman swarmed into the land. The English people grew into a powerful nation, and Nature still waited for a full return of the capital she had invested in the ancient club-mosses. The eighteenth century arrived, and with it James Watt. The brain of that man was the spore out of which was developed the modern steam-engine, and all the prodigious trees and branches of modern industry which have grown out of this. But coal is as much an essential condition of this growth and development as carbonic acid is for that of a club-moss. Wanting coal, we could not have smelted the iron needed to make our engines, nor have worked our engines when we had got them. But take away the engines, and the great towns of Yorkshire and Lancashire vanish like a dream. Manufactures give place to agriculture and pasture, and not ten men can live where now ten thousand are amply supported.

Thus, all this abundant wealth of money and of vivid life is Nature's interest upon her investment in club-mosses, and the like, so long ago. But what becomes of the coal which is burnt in yielding this interest? Heat comes out of it, light comes out of it; and if we could gather together all that goes up the chimney, and all that remains in the grate of a thoroughly-burnt coal-fire, we should find ourselves in possession of a quantity of carbonic acid, water, ammonia, and mineral matters, exactly equal in weight to the coal. But these are the very matters with which Nature supplied the club-mosses which made the coal

She is paid back principal and interest at the same time; and she straightway invests the carbonic acid, the water, and the ammonia in new forms of life, feeding with them the plants that now live. Thrifty Nature! Surely no prodigal, but most notable of housekeepers!

VI

ON THE BORDER TERRITORY BETWEEN THE ANIMAL AND THE VEGETABLE KINGDOMS

[1876]

IN the whole history of science there is nothing more remarkable than the rapidity of the growth of biological knowledge within the last half-century, and the extent of the modification which has thereby been effected in some of the fundamental conceptions of the naturalist.

In the second edition of the "Règne Animal," published in 1828, Cuvier devotes a special section to the "Division of Organised Beings into Animals and Vegetables," in which the question is treated with that comprehensiveness of knowledge and clear critical judgment which characterise his writings, and justify us in regarding them as representative expressions of the most extensive, if not the profoundest, knowledge of his time. He tells us that living beings have been subdivided from the earliest times into *animated beings*, which possess sense and motion, and *inanimated beings*, which are devoid of these functions and simply vegetate.

Although the roots of plants direct themselves towards moisture, and their leaves towards air and light,—although the parts of some plants exhibit oscillating movements without any perceptible cause, and the leaves of others retract when touched,—yet none of these movements justify the ascription to plants of perception or of will. From the mobility of animals, Cuvier, with his characteristic partiality for teleological reasoning, deduces the necessity of the existence in them of an alimentary cavity, or reservoir of food, whence their nutrition may be drawn by the vessels; which are a sort of internal roots; and, in the presence of this alimentary cavity, he naturally sees the primary and the most important distinction between animals and plants.

Following out his teleological argument, Cuvier remarks that the organisation of this cavity and its appurtenances must needs vary according to the nature of the aliment, and the operations which it has to undergo, before it can be converted into substances fitted for absorption; while the atmosphere and the earth supply plants with juices ready prepared, and which can be absorbed immediately. As the animal body required to be independent of heat and of the atmosphere, there

M 2

were no means by which the motion of its fluids
could be produced by internal causes. Hence
arose the second great distinctive character of
animals, or the circulatory system, which is less
important than the digestive, since it was unnecessary, and therefore is absent, in the more
simple animals.

Animals further needed muscles for locomotion and nerves for sensibility. Hence, says
Cuvier, it was necessary that the chemical composition of the animal body should be more complicated than that of the plant; and it is so, inasmuch
as an additional substance, nitrogen, enters into it
as an essential element; while, in plants, nitrogen
is only accidentally joined with the three other
fundamental constituents of organic beings—
carbon, hydrogen, and oxygen. Indeed, he afterwards affirms that nitrogen is peculiar to animals;
and herein he places the third distinction between
the animal and the plant. The soil and the
atmosphere supply plants with water, composed of
hydrogen and oxygen; air, consisting of nitrogen
and oxygen; and carbonic acid, containing carbon
and oxygen. They retain the hydrogen and the
carbon, exhale the superfluous oxygen, and absorb
little or no nitrogen. The essential character of
vegetable life is the exhalation of oxygen, which
is effected through the agency of light. Animals,
on the contrary, derive their nourishment either
directly or indirectly from plants. They get rid of

the superfluous hydrogen and carbon, and accumulate nitrogen. The relations of plants and animals to the atmosphere are therefore inverse. The plant withdraws water and carbonic acid from the atmosphere, the animal contributes both to it. Respiration—that is, the absorption of oxygen and the exhalation of carbonic acid—is the specially animal function of animals, and constitutes their fourth distinctive character.

Thus wrote Cuvier in 1828. But, in the fourth and fifth decades of this century, the greatest and most rapid revolution which biological science has ever undergone was effected by the application of the modern microscope to the investigation of organic structure; by the introduction of exact and easily manageable methods of conducting the chemical analysis of organic compounds; and finally, by the employment of instruments of precision for the measurement of the physical forces which are at work in the living economy.

That the semi-fluid contents (which we now term protoplasm) of the cells of certain plants, such as the *Charæ*, are in constant and regular motion, was made out by Bonaventura Corti a century ago; but the fact, important as it was, fell into oblivion, and had to be rediscovered by Treviranus in 1807. Robert Brown noted the more complex motions of the protoplasm in the cells of *Tradescantia* in 1831; and now such movements of the living substance of plants are well known to be some of the most widely-prevalent phenomena of vegetable life.

Agardh, and other of the botanists of Cuvier's generation, who occupied themselves with the lower plants, had observed that, under particular circumstances, the contents of the cells of certain water-weeds were set free, and moved about with considerable velocity, and with all the appearances of spontaneity, as locomotive bodies, which, from their similarity to animals of simple organisation, were called "zoospores." Even as late as 1845, however, a botanist of Schleiden's eminence dealt very sceptically with these statements; and his scepticism was the more justified, since Ehrenberg, in his elaborate and comprehensive work on the *Infusoria*, had declared the greater number of what are now recognised as locomotive plants to be animals.

At the present day, innumerable plants and free plant cells are known to pass the whole or part of their lives in an actively locomotive condition, in no wise distinguishable from that of one of the simpler animals; and, while in this condition, their movements are, to all appearance, as spontaneous —as much the product of volition—as those of such animals.

Hence the teleological argument for Cuvier's first diagnostic character—the presence in animals of an alimentary cavity, or internal pocket, in which they can carry about their nutriment—has broken down, so far, at least, as his mode of stating it goes. And, with the advance of microscopic anatomy, the universality of the fact itself among animals has ceased to be predicable. Many animals of even complex structure, which live parasitically within others, are wholly devoid of an alimentary cavity. Their food is provided for them, not only ready cooked, but ready digested, and the alimentary canal, become superfluous, has disappeared. Again, the males of most Rotifers have no digestive apparatus ; as a German naturalist has remarked, they devote themselves entirely to the " Minnedienst," and are to be reckoned among the few realisations of the Byronic ideal of a lover. Finally, amidst the lowest forms of animal life, the speck of gelatinous protoplasm, which constitutes the whole body, has no permanent digestive cavity or mouth, but takes in its food anywhere ; and digests, so to speak, all over its body.

But although Cuvier's leading diagnosis of the animal from the plant will not stand a strict test, it remains one of the most constant of the distinctive characters of animals. And, if we substitute for the possession of an alimentary cavity, the power of taking solid nutriment into the body and there digesting it, the definition so changed will cover all animals, except certain parasites, and the few and exceptional cases of non-parasitic animals which do not feed at all. On the other hand, the definition thus amended will exclude all ordinary vegetable organisms.

Cuvier himself practically gives up his second distinctive mark when he admits that it is wanting in the simpler animals.

The third distinction is based on a completely erroneous conception of the chemical differences and resemblances between the constituents of animal and vegetable organisms, for which Cuvier is not responsible, as it was current among contemporary chemists. It is now established that nitrogen is as essential a constituent of vegetable as of animal living matter; and that the latter is, chemically speaking, just as complicated as the former. Starchy substances, cellulose and sugar, once supposed to be exclusively confined to plants, are now known to be regular and normal products of animals. Amylaceous and saccharine substances are largely manufactured, even by the highest animals; cellulose is widespread as a constituent of the skeletons of the lower animals; and it is probable that amyloid substances are universally present in the animal organism, though not in the precise form of starch.

Moreover, although it remains true that there is an inverse relation between the green plant in sunshine and the animal, in so far as, under these circumstances, the green plant decomposes carbonic acid and exhales oxygen, while the animal absorbs oxygen and exhales carbonic acid; yet,

the exact researches of the modern chemical investigators of the physiological processes of plants
have clearly demonstrated the fallacy of attempting to draw any general distinction between
animals and vegetables on this ground.   In fact,
the difference vanishes with the sunshine, even in
the case of the green plant; which, in the dark,
absorbs oxygen and gives out carbonic acid like
any animal.[1]   On the other hand, those plants,
such as the fungi, which contain no chlorophyll
and are not green, are always, so far as respiration
is concerned, in the exact position of animals.
They absorb oxygen and give out carbonic acid.

Thus, by the progress of knowledge, Cuvier's
fourth distinction between the animal and the
plant has been as completely invalidated as the
third and second; and even the first can be retained only in a modified form and subject to
exceptions.

But has the advance of biology simply tended
to break down old distinctions, without establishing new ones?

With a qualification, to be considered presently,
the answer to this question is undoubtedly in the
affirmative.   The famous researches of Schwann

[1] There is every reason to believe that living plants, like living
animals, always respire, and, in respiring, absorb oxygen and
give off carbonic acid; but, that in green plants exposed
to daylight or to the electric light, the quantity of oxygen
evolved in consequence of the decomposition of carbonic acid
by a special apparatus which green plants possess exceeds that
absorbed in the concurrent respiratory process.

and Schleiden in 1837 and the following years, founded the modern science of histology, or that branch of anatomy which deals with the ultimate visible structure of organisms, as revealed by the microscope ; and, from that day to this, the rapid improvement of methods of investigation, and the energy of a host of accurate observers, have given greater and greater breadth and firmness to Schwann's great generalisation, that a fundamental unity of structure obtains in animals and plants ; and that, however diverse may be the fabrics, or *tissues*, of which their bodies are composed, all these varied structures result from the metamorphosis of morphological units (termed *cells*, in a more general sense than that in which the word " cells " was at first employed), which are not only similar in animals and in plants respectively, but present a close resemblance, when those of animals and those of plants are compared together.

The contractility which is the fundamental condition of locomotion, has not only been discovered to exist far more widely among plants than was formerly imagined ; but, in plants, the act of contraction has been found to be accompanied, as Dr. Burdon Sanderson's interesting investigations have shown, by a disturbance of the electrical state of the contractile substance, comparable to that which was found by Du Bois Reymond to be a concomitant of the activity of ordinary muscle in animals.

Again, I know of no test by which the reaction of the leaves of the Sundew and of other plants to stimuli, so fully and carefully studied by Mr. Darwin, can be distinguished from those acts of contraction following upon stimuli, which are called " reflex " in animals.

On each lobe of the bilobed leaf of Venus's fly-trap (*Dionæa muscipula*) are three delicate filaments which stand out at right angle from the surface of the leaf. Touch one of them with the end of a fine human hair and the lobes of the leaf instantly close together [1] in virtue of an act of contraction of part of their substance, just as the body of a snail contracts into its shell when one of its " horns " is irritated.

The reflex action of the snail is the result of the presence of a nervous system in the animal. A molecular change takes place in the nerve of the tentacle, is propagated to the muscles by which the body is retracted, and causing them to contract, the act of retraction is brought about. Of course the similarity of the acts does not necessarily involve the conclusion that the mechanism by which they are effected is the same; but it suggests a suspicion of their identity which needs careful testing.

The results of recent inquiries into the structure of the nervous system of animals converge towards the conclusion that the nerve fibres, which we

---

[1] Darwin, *Insectivorous Plants*, p. 289.

have hitherto regarded as ultimate elements of
nervous tissue, are not such, but are simply the
visible aggregations of vastly more attenuated
filaments, the diameter of which dwindles down to
the limits of our present microscopic vision, greatly
as these have been extended by modern improve-
ments of the microscope; and that a nerve is, in
its essence, nothing but a linear tract of specially
modified protoplasm between two points of an
organism—one of which is able to affect the other
by means of the communication so established.
Hence, it is conceivable that even the simplest
living being may possess a nervous system. And
the question whether plants are provided with a
nervous system or not, thus acquires a new aspect,
and presents the histologist and physiologist with
a problem of extreme difficulty, which must be
attacked from a new point of view and by the aid
of methods which have yet to be invented.

Thus it must be admitted that plants may be
contractile and locomotive; that, while locomotive,
their movements may have as much appearance of
spontaneity as those of the lowest animals; and
that many exhibit actions, comparable to those
which are brought about by the agency of a
nervous system in animals. And it must be
allowed to be possible that further research may
reveal the existence of something comparable to a
nervous system in plants. So that I know not
where we can hope to find any absolute distinction

between animals and plants, unless we return to their mode of nutrition, and inquire whether certain differences of a more occult character than those imagined to exist by Cuvier, and which certainly hold good for the vast majority of animals and plants, are of universal application.

A bean may be supplied with water in which salts of ammonia and certain other mineral salts are dissolved in due proportion; with atmospheric air containing its ordinary minute dose of carbonic acid; and with nothing else but sunlight and heat. Under these circumstances, unnatural as they are, with proper management, the bean will thrust forth its radicle and its plumule; the former will grow down into roots, the latter grow up into the stem and leaves of a vigorous bean-plant; and this plant will, in due time, flower and produce its crop of beans, just as if it were grown in the garden or in the field.

The weight of the nitrogenous protein compounds, of the oily, starchy, saccharine and woody substances contained in the full-grown plant and its seeds, will be vastly greater than the weight of the same substances contained in the bean from which it sprang. But nothing has been supplied to the bean save water, carbonic acid, ammonia, potash, lime, iron, and the like, in combination with phosphoric, sulphuric, and other acids. Neither protein, nor fat, nor starch, nor sugar, nor any substance in the slightest degree resembling them, has formed part of the food of the bean. But the weights of the carbon, hydrogen, oxygen, nitrogen, phosphorus, sulphur, and other elementary bodies contained in the bean-plant, and in the seeds which it produces, are exactly equivalent to the weights of the same elements which have disappeared from the materials supplied to the bean during its growth. Whence it follows that the bean has taken in only the raw materials of its fabric, and has manufactured them into bean-stuffs.

The bean has been able to perform this great chemical feat by the help of its green colouring matter, or chlorophyll; for it is only the green parts of the plant which, under the influence of sunlight, have the marvellous power of decomposing carbonic acid, setting free the oxygen and laying hold of the carbon which it contains. In fact, the bean obtains two of the absolutely indispensable elements .of its substance from two distinct sources; the watery solution, in which its roots are plunged, contains nitrogen but no carbon ; the air, to which the leaves are exposed, contains carbon, but its nitrogen is in the state of a free gas, in which condition the bean can make no use of it ; [1] and the chlorophyll [2] is the apparatus by

[1] I purposely assume that the air with which the bean is supplied in the case stated contains no ammoniacal salts.

[2] The recent researches of Pringsheim have raised a host of questions as to the exact share taken by chlorophyll in the chemical operations which are effected by the green parts of plants. It may be that the chlorophyll is only a constant concomitant of the actual deoxidising apparatus.

which the carbon is extracted from the atmospheric carbonic acid—the leaves being the chief
laboratories in which this operation is effected.

The great majority of conspicuous plants are, as
everybody knows, green; and this arises from the
abundance of their chlorophyll. The few which
contain no chlorophyll and are colourless, are unable to extract the carbon which they require from
atmospheric carbonic acid, and lead a parasitic
existence upon other plants; but it by no means
follows, often as the statement has been repeated,
that the manufacturing power of plants depends
on their chlorophyll, and its interaction with the
rays of the sun. On the contrary, it is easily
demonstrated, as Pasteur first proved, that the
lowest fungi, devoid of chlorophyll, or of any substitute for it, as they are, nevertheless possess the
characteristic manufacturing powers of plants in a
very high degree. Only it is necessary that they
should be supplied with a different kind of raw
material; as they cannot extract carbon from carbonic acid, they must be furnished with something
else that contains carbon. Tartaric acid is such a
substance; and if a single spore of the commonest
and most troublesome of moulds—*Penicillium*—be
sown in a saucerful of water, in which tartrate of
ammonia, with a small percentage of phosphates
and sulphates is contained, and kept warm, whether
in the dark or exposed to light, it will, in a
short time, give rise to a thick crust of mould,

which contains many million times the weight of the original spore, in protein compounds and cellulose. Thus we have a very wide basis of fact for the generalisation that plants are essentially characterised by their manufacturing capacity—by their power of working up mere mineral matters into complex organic compounds.

Contrariwise, there is a no less wide foundation for the generalisation that animals, as Cuvier puts it, depend directly or indirectly upon plants for the materials of their bodies; that is, either they are herbivorous, or they eat other animals which are herbivorous.

But for what constituents of their bodies are animals thus dependent upon plants? Certainly not for their horny matter; nor for chondrin, the proximate chemical element of cartilage; nor for gelatine; nor for syntonin, the constituent of muscle; nor for their nervous or biliary substances; nor for their amyloid matters; nor, necessarily, for their fats.

It can be experimentally demonstrated that animals can make these for themselves. But that which they cannot make, but must, in all known cases, obtain directly or indirectly from plants, is the peculiar nitrogenous matter, protein. Thus the plant is the ideal *prolétaire* of the living world, the worker who produces; the animal, the ideal aristocrat, who mostly occupies himself in consuming, after the manner of that noble repre-

sentative of the line of Zähdarm, whose epitaph is
written in " Sartor Resartus."

Here is our last hope of finding a sharp line of
demarcation between plants and animals; for, as
I have already hinted, there is a border territory
between the two kingdoms, a sort of no-man's-
land, the inhabitants of which certainly cannot
be discriminated and brought to their proper
allegiance in any other way.

Some months ago, Professor Tyndall asked me
to examine a drop of infusion of hay, placed
under an excellent and powerful microscope, and
to tell him what I thought some organisms
visible in it were. I looked and observed, in the
first place, multitudes of *Bacteria* moving about
with their ordinary intermittent spasmodic
wriggles. As to the vegetable nature of these
there is now no doubt. Not only does the close
resemblance of the *Bacteria* to unquestionable
plants, such as the *Oscillatoriæ* and the lower forms
of *Fungi*, justify this conclusion, but the manufacturing test settles the question at once. It
is only needful to add a minute drop of fluid
containing *Bacteria*, to water in which tartrate,
phosphate, and sulphate of ammonia are dissolved;
and, in a very short space of time, the clear fluid
becomes milky by reason of their prodigious
multiplication, which, of course, implies the
manufacture of living Bacterium-stuff out of
these merely saline matters.

But other active organisms, very much larger than the *Bacteria*, attaining in fact the comparatively gigantic dimensions of $\frac{1}{3000}$ of an inch or more, incessantly crossed the field of view. Each of these had a body shaped like a pear, the small end being slightly incurved and produced into a long curved filament, or *cilium*, of extreme tenuity. Behind this, from the concave side of the incurvation, proceeded another long cilium, so delicate as to be discernible only by the use of the highest powers and careful management of the light. In the centre of the pear-shaped body a clear round space could occasionally be discerned, but not always; and careful watching showed that this clear vacuity appeared gradually, and then shut up and disappeared suddenly, at regular intervals. Such a structure is of common occurrence among the lowest plants and animals, and is known as a *contractile vacuole*.

The little creature thus described sometimes propelled itself with great activity, with a curious rolling motion, by the lashing of the front cilium, while the second cilium trailed behind; sometimes it anchored itself by the hinder cilium and was spun round by the working of the other, its motions resembling those of an anchor buoy in a heavy sea. Sometimes, when two were in full career towards one another, each would appear dexterously to get out of the other's way; sometimes a crowd would assemble and jostle one

another, with as much semblance of individual
effort as a spectator on the Grands Mulets might
observe with a telescope among the specks representing men in the valley of Chamounix.

The spectacle, though always surprising, was
not new to me. So my reply to the question put
to me was, that these organisms were what
biologists call *Monads*, and though they might be
animals, it was also possible that they might,
like the *Bacteria*, be plants. My friend received
my verdict with an expression which showed a
sad want of respect for authority. He would as
soon believe that a sheep was a plant. Naturally
piqued by this want of faith, I have thought a
good deal over the matter ; and, as I still rest in
the lame conclusion I originally expressed, and
must even now confess that I cannot certainly
say whether this creature is an animal or a plant,
I think it may be well to state the grounds of my
hesitation at length. But, in the first place, in
order that I may conveniently distinguish this
"Monad" from the multitude of other things
which go by the same designation, I must give it
a name of its own. I think (though, for reasons
which need not be stated at present, I am not
quite sure) that it is identical with the species
Monas lens, as defined by the eminent French
microscopist Dujardin, though his magnifying
power was probably insufficient to enable him
to see that it is curiously like a much larger

N 2

form of monad which he has named *Heteromita*. I shall, therefore, call it not *Monas*, but *Heteromita lens*.

I have been unable to devote to my *Heteromita* the prolonged study needful to work out its whole history, which would involve weeks, or it may be months, of unremitting attention. But I the less regret this circumstance, as some remarkable observations recently published by Messrs. Dallinger and Drysdale[1] on certain Monads, relate, in part, to a form so similar to my *Heteromita lens*, that the history of the one may be used to illustrate that of the other. These most patient and painstaking observers, who employed the highest attainable powers of the microscope and, relieving one another, kept watch day and night over the same individual monads, have been enabled to trace out the whole history of their *Heteromita*; which they found in infusions of the heads of fishes of the Cod tribe.

Of the four monads described and figured by these investigators, one, as I have said, very closely resembles *Heteromita lens* in every particular, except that it has a separately distinguishable central particle or " nucleus," which is not certainly to be made out in *Heteromita lens*; and that nothing is said by Messrs. Dallinger

---

[1] " Researches in the Life-history of a Cercomonad : a Lesson in Biogenesis"; and " Further Researches in the Life-history of the Monads."—*Monthly Microscopical Journal*, 1873.

and Drysdale of the existence of a contractile vacuole in this monad, though they describe it in another.

Their *Heteromita*, however, multiplied rapidly by fission. Sometimes a transverse constriction appeared; the hinder half developed a new cilium, and the hinder cilium gradually split from its base to its free end, until it was divided into two; a process which, considering the fact that this fine filament cannot be much more than $\frac{1}{100000}$ of an inch in diameter, is wonderful enough. The constriction of the body extended inwards until the two portions were united by a narrow isthmus; finally, they separated and each swam away by itself, a complete *Heteromita*, provided with its two cilia. Sometimes the constriction took a longitudinal direction, with the same ultimate result. In each case the process occupied not more than six or seven minutes. At this rate, a single *Heteromita* would give rise to a thousand like itself in the course of an hour, to about a million in two hours, and to a number greater than the generally assumed number of human beings now living in the world in three hours; or, if we give each *Heteromita* an hour's enjoyment of individual existence, the same result will be obtained in about a day. The apparent suddenness of the appearance of multitudes of such organisms as these, in any nutritive fluid to which one obtains access, is thus easily explained.

During these processes of multiplication by fission, the *Heteromita* remains active; but sometimes another mode of fission occurs. The body becomes rounded and quiescent, or nearly so; and, while in this resting state, divides into two portions, each of which is rapidly converted into an active *Heteromita*.

A still more remarkable phenomenon is that kind of multiplication which is preceded by the union of two monads, by a process which is termed *conjugation*. Two active *Heteromitæ* become applied to one another, and then slowly and gradually coalesce into one body. The two nuclei run into one; and the mass resulting from the conjugation of the two *Heteromitæ*, thus fused together, has a triangular form. The two pairs of cilia are to be seen, for some time, at two of the angles, which answer to the small ends of the conjoined monads; but they ultimately vanish, and the twin organism, in which all visible traces of organisation have disappeared, falls into a state of rest. Sudden wave-like movements of its substance next occur; and, in a short time, the apices of the triangular mass burst, and give exit to a dense yellowish, glairy fluid, filled with minute granules. This process, which, it will be observed, involves the actual confluence and mixture of the substance of two distinct organisms, is effected in the space of about two hours.

The authors whom I quote say that they

"cannot express" the excessive minuteness of the granules in question, and they estimate their diameter at less than $\frac{1}{200000}$ of an inch. Under the highest powers of the microscope, at present applicable, such specks are hardly discernible. Nevertheless, particles of this size are massive when compared to physical molecules; whence there is no reason to doubt that each, small as it is, may have a molecular structure sufficiently complex to give rise to the phenomena of life. And, as a matter of fact, by patient watching of the place at which these infinitesimal living particles were discharged, our observers assured themselves of their growth and development into new monads. In about four hours from their being set free, they had attained a sixth of the length of the parent, with the characteristic cilia, though at first they were quite motionless; and, in four hours more, they had attained the dimensions and exhibited all the activity of the adult. These inconceivably minute particles are therefore the germs of the *Heteromita*; and from the dimensions of these germs it is easily shown that the body formed by conjugation may, at a low estimate, have given exit to thirty thousand of them; a result of a matrimonial process whereby the contracting parties, without a metaphor, "become one flesh," enough to make a Malthusian despair of the future of the Universe.

I am not aware that the investigators from

whom I have borrowed this history have endeavoured to ascertain whether their monads take solid nutriment or not; so that though they help us very much to fill up the blanks in the history of my *Heteromita*, their observations throw no light on the problem we are trying to solve—Is it an animal or is it a plant?

Undoubtedly it is possible to bring forward very strong arguments in favour of regarding *Heteromita* as a plant.

For example, there is a Fungus, an obscure and almost microscopic mould, termed *Peronospora infestans*. Like many other Fungi, the *Peronosporæ* are parasitic upon other plants; and this particular *Peronospora* happens to have attained much notoriety and political importance, in a way not without a parallel in the career of notorious politicians, namely, by reason of the frightful mischief it has done to mankind. For it is this *Fungus* which is the cause of the potato disease; and, therefore, *Peronospora infestans* (doubtless of exclusively Saxon origin, though not accurately known to be so) brought about the Irish famine. The plants afflicted with the malady are found to be infested by a mould, consisting of fine tubular filaments, termed *hyphæ*, which burrow through the substance of the potato plant, and appropriate to themselves the substance of their host; while, at the same time, directly or indirectly, they set up chemical changes by which even its woody framework becomes blackened, sodden, and withered.

In structure, however, the *Peronospora* is as much a mould as the common *Penicillium*; and just as the *Penicillium* multiplies by the breaking up of its hyphæ into separate rounded bodies, the spores; so, in the *Peronospora*, certain of the hyphæ grow out into the air through the interstices of the superficial cells of the potato plant, and develop spores. Each of these hyphæ usually gives off several branches. The ends of the branches dilate and become closed sacs, which eventually drop off as spores. The spores falling on some part of the same potato plant, or carried by the wind to another, may at once germinate, throwing out tubular prolongations which become hyphæ, and burrow into the substance of the plant attacked. But, more commonly, the contents of the spore divide into six or eight separate portions. The coat of the spore gives way, and each portion then emerges as an independent organism, which has the shape of a bean, rather narrower at one end than the other, convex on one side, and depressed or concave on the opposite. From the depression, two long and delicate cilia proceed, one shorter than the other, and directed forwards. Close to the origin of these cilia, in the substance of the body, is a regularly pulsating, contractile vacuole. The shorter cilium vibrates actively, and effects the locomotion of the organism, while the other trails behind; the whole body rolling on its axis with its pointed end forwards.

The eminent botanist, De Bary, who was not thinking of our problem, tells us, in describing the movements of these "Zoospores," that, as they swim about, " Foreign bodies are carefully avoided, and the whole movement has a deceptive likeness to the voluntary changes of place which are observed in microscopic animals."

After swarming about in this way in the moisture on the surface of a leaf or stem (which, film though it may be, is an ocean to such a fish) for half an hour, more or less, the movement of the zoospore becomes slower, and is limited to a slow turning upon its axis, without change of place. It then becomes quite quiet, the cilia disappear, it assumes a spherical form, and surrounds itself with a distinct, though delicate, membranous coat. A protuberance then grows out from one side of the sphere, and rapidly increasing in length, assumes the character of a hypha. The latter penetrates into the substance of the potato plant, either by entering a stomate, or by boring through the wall of an epidermic cell, and ramifies, as a mycelium, in the substance of the plant, destroying the tissues with which it comes in contact. As these processes of multiplication take place very rapidly, millions of spores are soon set free from a single infested plant; and, from their minuteness,

they are readily transported by the gentlest
breeze. Since, again, the zoospores set free from
each spore, in virtue of their powers of locomotion,
swiftly disperse themselves over the surface, it is
no wonder that the infection, once started, soon
spreads from field to field, and extends its ravages
over a whole country.

However, it does not enter into my present
plan to treat of the potato disease, instructively as
its history bears upon that of other epidemics;
and I have selected the case of the *Peronospora*
simply because it affords an example of an organism, which, in one stage of its existence, is truly a
" Monad," indistinguishable by any important
character from our *Heteromita*, and extraordinarily
like it in some respects. And yet this " Monad "
can be traced, step by step, through the series of
metamorphoses which I have described, until it
assumes the features of an organism, which is as
much a plant as is an oak or an elm.

Moreover, it would be possible to pursue the
analogy farther. Under certain circumstances, a
process of conjugation takes place in the *Peronospora*. Two separate portions of its protoplasm
become fused together, surround themselves with
a thick coat, and give rise to a sort of vegetable
egg called an *oospore*. After a period of rest, the
contents of the oospore break up into a number of
zoospores like those already described, each of
which, after a period of activity, germinates in the

ordinary way.   This process obviously corresponds
with the conjugation and subsequent setting free
of germs in the *Heteromita*.

But it may be said that the *Peronospora* is,
after all, a questionable sort of plant; that it seems
to be wanting in the manufacturing power, selected
as the main distinctive character of vegetable
life; or, at any rate, that there is no proof that
it does not get its protein matter ready made
from the potato plant.

Let us, therefore, take a case which is not open
to these objections.

There are some small plants known to botanists
as members of the genus *Coleochæte*, which, without being truly parasitic, grow upon certain
water-weeds, as lichens grow upon trees.   The
little plant has the form of an elegant green star,
the branching arms of which are divided into
cells.   Its greenness is due to its chlorophyll, and
it undoubtedly has the manufacturing power in
full degree, decomposing carbonic acid and setting
oxygen free, under the influence of sunlight.   But
the protoplasmic contents of some of the cells of
which the plant is made up occasionally divide, by
a method similar to that which effects the division
of the contents of the *Peronospora* spore; and the
severed portions are then set free as active monad-
like zoospores.   Each is oval and is provided at
one extremity with two long active cilia.   Propelled by these, it swims about for a longer or

shorter time, but at length comes to a state of
rest and gradually grows into a *Coleochæte*.
Moreover, as in the *Peronospora*, conjugation may
take place and result in an oospore; the contents
of which divide and are set free as monadiform
germs.

If the whole history of the zoospores of *Peronospora* and of *Coleochæte* were unknown, they would
undoubtedly be classed among "Monads" with
the same right as *Heteromita*; why then may not
Heteromita be a plant, even though the cycle of
forms through which it passes shows no terms
quite so complex as those which occur in *Peronospora* and *Coleochæte?* And, in fact, there are
some green organisms, in every respect characteristically plants, such as *Chlamydomonas*,
and the common *Volvox*, or so-called "Globe
animalcule," which run through a cycle of forms
of just the same simple character as those of
Heteromita.

The name of *Chlamydomonas* is applied to certain
microscopic green bodies, each of which consists of
a protoplasmic central substance invested by a
structureless sac. The latter contains cellulose, as
in ordinary plants; and the chlorophyll which
gives the green colour enables the *Chlamydomonas*
to decompose carbonic acid and fix carbon as they
do. Two long cilia protrude through the cell-wall,
and effect the rapid locomotion of this "monad,"
which, in all respects except its mobility, is

characteristically a plant. Under ordinary circumstances, the *Chlamydomonas* multiplies by
simple fission, each splitting into two or into four
parts, which separate and become independent
organisms. Sometimes, however, the *Chlamydomonas* divides into eight parts, each of which is
provided with four instead of two cilia. These
"zoospores" conjugate in pairs, and give rise to
quiescent bodies, which multiply by division, and
eventually pass into the active state.

Thus, so far as outward form and the general
character of the cycle of modifications, through
which the organism passes in the course of its
life, are concerned, the resemblance between
Chlamydomonas and *Heteromita* is of the closest
description. And on the face of the matter there
is no ground for refusing to admit that *Heteromita*
may be related to *Chlamydomonas*, as the colourless
fungus is to the green alga. *Volvox* may be compared to a hollow sphere, the wall of which is
made up of coherent Chlamydomonads; and which
progresses with a rotating motion effected by the
paddling of the multitudinous pairs of cilia which
project from its surface. Each *Volvox*-monad,
moreover, possesses a red pigment spot, like the
simplest form of eye known among animals. The
methods of fissive multiplication and of conjugation
observed in the monads of this locomotive globe
are essentially similar to those observed in *Chlamydomonas*; and, though a hard battle has been

fought over it, *Volvox* is now finally surrendered to the Botanists.

Thus there is really no reason why *Heteromita* may not be a plant; and this conclusion would be very satisfactory, if it were not equally easy to show that there is really no reason why it should not be an animal. For there are numerous organisms presenting the closest resemblance to *Heteromita*, and, like it, grouped under the general name of " Monads," which, nevertheless, can be observed to take in solid nutriment, and which, therefore, have a virtual, if not an actual, mouth and digestive cavity, and thus come under Cuvier's definition of an animal. Numerous forms of such animals have been described by Ehrenberg, Dujardin, H. James Clark, and other writers on the *Infusoria*. Indeed, in another infusion of hay in which my *Heteromita lens* occurred, there were innumerable such infusorial animalcules belonging to the well-known species *Colpoda cucullus*.[1]

Full-sized specimens of this animalcule attain a length of between $\frac{1}{300}$ or $\frac{1}{400}$ of an inch, so that it may have ten times the length and a thousand times the mass of a *Heteromita*. In shape, it is not altogether unlike *Heteromita*. The small end, however, is not produced into one long cilium, but the general surface of the body is covered with

---

[1] Excellently described by Stein, almost all of whose statements I have verified.

small actively vibrating ciliary organs, which are only longest at the small end. At the point which answers to that from which the two cilia arise in *Heteromita*, there is a conical depression, the mouth; and, in young specimens, a tapering filament, which reminds one of the posterior cilium of *Heteromita*, projects from this region.

The body consists of a soft granular proto-plasmic substance, the middle of which is occupied by a large oval mass called the " nucleus "; while, at its hinder end, is a " contractile vacuole," con-spicuous by its regular rhythmic appearances and disappearances. Obviously, although the *Colpoda* is not a monad, it differs from one only in subor-dinate details. Moreover, under certain conditions, it becomes quiescent, incloses itself in a delicate case or *cyst*, and then divides into two, four, or more portions, which are eventually set free and swim about as active *Colpodæ*.

But this creature is an unmistakable animal, and full-sized *Colpodæ* may be fed as easily as one feeds chickens. It is only needful to diffuse very finely ground carmine through the water in which they live, and, in a very short time, the bodies of the *Colpodæ* are stuffed with the deeply-coloured granules of the pigment.

And if this were not sufficient evidence of the animality of *Colpoda*, there comes the fact that it is even more similar to another well-knowm animalcule, *Paramœcium*, than it is to a monad.

But *Paramœcium* is so huge a creature compared
with those hitherto discussed—it reaches $\frac{1}{120}$ of
an inch or more in length—that there is no difficulty in making out its organisation in detail;
and in proving that it is not only an animal, but
that it is an animal which possesses a somewhat
complicated organisation. For example, the surface layer of its body is different in structure from
the deeper parts. There are two contractile
vacuoles, from each of which radiates a system of
vessel-like canals; and not only is there a conical
depression continuous with a tube, which serve as
mouth and gullet, but the food ingested takes a
definite course, and refuse is rejected from a
definite region. Nothing is easier than to feed
these animals, and to watch the particles of indigo
or carmine accumulate at the lower end of the
gullet. From this they gradually project, surrounded by a ball of water, which at length passes
with a jerk, oddly simulating a gulp, into the
pulpy central substance of the body, there to cir-
culate up one side and down the other, until its
contents are digested and assimilated. Nevertheless, this complex animal multiplies by division, as
the monad does, and, like the monad, undergoes
conjugation. It stands in the same relation to
Heteromita on the animal side, as *Coleochœte* does
on the plant side. Start from either, and such an
insensible series of gradations leads to the monad
that it is impossible to say at any stage of the

progress where the line between the animal and the plant must be drawn.

There is reason to think that certain organisms which pass through a monad stage of existence, such as the *Myxomycetes*, are, at one time of their lives, dependent upon external sources for their protein matter, or are animals; and, at another period, manufacture it, or are plants. And seeing that the whole progress of modern investigation is in favour of the doctrine of continuity, it is a fair and probable speculation—though only a speculation—that, as there are some plants which can manufacture protein out of such apparently intractable mineral matters as carbonic acid, water, nitrate of ammonia, metallic and earthy salts; while others need to be supplied with their carbon and nitrogen in the somewhat less raw form of tartrate of ammonia and allied compounds; so there may be yet others, as is possibly the case with the true parasitic plants, which can only manage to put together materials still better prepared—still more nearly approximated to protein—until we arrive at such organisms as the *Psorospermiæ* and the *Panhistophyton*, which are as much animal as vegetable in structure, but are animal in their dependence on other organisms for their food.

The singular circumstance observed by Meyer, that the *Torula* of yeast, though an indubitable plant, still flourishes most vigorously when supplied with the complex nitrogenous substance, pepsin;

the probability that the *Peronospora* is nourished directly by the protoplasm of the potato-plant; and the wonderful facts which have recently been brought to light respecting insectivorous plants, all favour this view; and tend to the conclusion that the difference between animal and plant is one of degree rather than of kind, and that the problem whether, in a given case, an organism is an animal or a plant, may be essentially insoluble.

VII

A LOBSTER; OR, THE STUDY OF ZOOLOGY

[1861]

NATURAL HISTORY is the name familiarly applied to the study of the properties of such natural bodies as minerals, plants, and animals; the sciences which embody the knowledge man has acquired upon these subjects are commonly termed Natural Sciences, in contradistinction to other so-called "physical" sciences; and those who devote themselves especially to the pursuit of such sciences have been and are commonly termed "Naturalists."

Linnæus was a naturalist in this wide sense, and his "Systema Naturæ" was a work upon natural history, in the broadest acceptation of the term; in it, that great methodising spirit embodied all that was known in his time of the distinctive characters of minerals, animals, and plants. But the enormous stimulus which Linnæus gave to the investigation of nature soon rendered it impossible that any one man should write another "Systema Naturæ," and extremely difficult for any one to become even a naturalist such as Linnæus was.

Great as have been the advances made by all the three branches of science, of old included under the title of natural history, there can be no doubt that zoology and botany have grown in an enormously greater ratio than mineralogy; and hence, as I suppose, the name of "natural history" has gradually become more and more definitely attached to these prominent divisions of the subject, and by "naturalist" people have meant more and more distinctly to imply a student of the structure and function of living beings.

However this may be, it is certain that the advance of knowledge has gradually widened the distance between mineralogy and its old associates, while it has drawn zoology and botany closer together; so that of late years it has been found convenient (and indeed necessary) to associate the sciences which deal with vitality and all its phenomena under the common head of "biology"; and the biologists have come to repudiate any blood-relationship with their foster-brothers, the mineralogists.

Certain broad laws have a general application throughout both the animal and the vegetable worlds, but the ground common to these kingdoms of nature is not of very wide extent, and the multiplicity of details is so great, that the student of living beings finds himself obliged to devote his attention exclusively either to the one or the other. If he elects to study plants, under any aspect, we know at once what to call him. He is a botanist, and his science is botany. But if the investigation of animal life be his choice, the name generally applied to him will vary according to the kind of animals he studies, or the particular phenomena of animal life to which he confines his attention. If the study of man is his object, he is called an anatomist, or a physiologist, or an ethnologist ; but if he dissects animals, or examines into the mode in which their functions are performed, he is a comparative anatomist or comparative physiologist. If he turns his attention to fossil animals, he is a palæontologist. If his mind is more particularly directed to the specific description, discrimination, classification, and distribution of animals, he is termed a zoologist.

For the purpose of the present discourse, however, I shall recognise none of these titles save the last, which I shall employ as the equivalent of botanist, and I shall use the term zoology as denoting the whole doctrine of animal life, in contradistinction to botany, which signifies the whole doctrine of vegetable life.

Employed in this sense, zoology, like botany, is

divisible into three great but subordinate sciences, morphology, physiology, and distribution, each of which may, to a very great extent, be studied independently of the other.

Zoological morphology is the doctrine of animal form or structure. Anatomy is one of its branches; development is another; while classification is the expression of the relations which different animals bear to one another, in respect of their anatomy and their development.

Zoological distribution is the study of animals in relation to the terrestrial conditions which obtain now, or have obtained at any previous epoch of the earth's history.

Zoological physiology, lastly, is the doctrine of the functions or actions of animals. It regards animal bodies as machines impelled by certain forces, and performing an amount of work which can be expressed in terms of the ordinary forces of nature. The final object of physiology is to deduce the facts of morphology, on the one hand, and those of distribution on the other, from the laws of the molecular forces of matter.

Such is the scope of zoology. But if I were to content myself with the enunciation of these dry definitions, I should ill exemplify that method of teaching this branch of physical science, which it is my chief business to-night to recommend. Let us turn away then from abstract definitions. Let us take some concrete living thing, some animal, the

commoner the better, and let us see how the application of common sense and common logic to the
obvious facts it presents, inevitably leads us into
all these branches of zoological science.

I have before me a lobster. When I examine
it, what appears to be the most striking character it
presents? Why, I observe that this part which we
call the tail of the lobster, is made up of six distinct
hard rings and a seventh terminal piece. If I
separate one of the middle rings, say the third, I
find it carries upon its under surface a pair of
limbs or appendages, each of which consists of a
stalk and two terminal pieces. So that I can represent a transverse section of the ring and its
appendages upon the diagram board in this way.

If I now take the fourth ring, I find it has the
same structure, and so have the fifth and the second ;
so that, in each of these divisions of the tail, I find
parts which correspond with one another, a ring
and two appendages ; and in each appendage a
stalk and two end pieces. These corresponding
parts are called, in the technical language of
anatomy, " homologous parts." The ring of the
third division is the " homologue " of the ring of
the fifth, the appendage of the former is the homologue of the appendage of the latter. And, as
each division exhibits corresponding parts in
corresponding places, we say that all the divisions
are constructed upon the same plan. But now let
us consider the sixth division. It is similar to,

and yet different from, the others. The ring is
essentially the same as in the other divisions ; but
the appendages look at first as if they were very
different ; and yet when we regard them closely,
what do we find ? A stalk and two terminal
divisions, exactly as in the others, but the stalk is
very short and very thick, the terminal divisions
are very broad and flat, and one of them is divided
into two pieces.

I may say, therefore, that the sixth segment is
like the others in plan, but that it is modified in
its details.

The first segment is like the others, so far as its
ring is concerned, and though its appendages differ
from any of those yet examined in the simplicity
of their structure, parts corresponding with the
stem and one of the divisions of the appendages
of the other segments can be readily discerned in
them.

Thus it appears that the lobster's tail is composed of a series of segments which are funda-
mentally similar, though each presents peculiar
modifications of the plan common to all. But
when I turn to the forepart of the body I see, at
first, nothing but a great shield-like shell, called
technically the " carapace," ending in front in a
sharp spine, on either side of which are the curious
compound eyes, set upon the ends of stout movable
stalks. Behind these, on the under side of the
body, are two pairs of long feelers, or antennæ,

followed by six pairs of jaws folded against one another over the mouth, and five pairs of legs, the foremost of these being the great pinchers, or claws, of the lobster.

It looks, at first, a little hopeless to attempt to find in this complex mass a series of rings, each with its pair of appendages, such as I have shown you in the abdomen, and yet it is not difficult to demonstrate their existence. Strip off the legs, and you will find that each pair is attached to a very definite segment of the under wall of the body; but these segments, instead of being the lower parts of free rings, as in the tail, are such parts of rings which are all solidly united and bound together; and the like is true of the jaws, the feelers, and the eye-stalks, every pair of which is borne upon its own special segment. Thus the conclusion is gradually forced upon us, that the body of the lobster is composed of as many rings as there are pairs of appendages, namely, twenty in all, but that the six hindmost rings remain free and movable, while the fourteen front rings become firmly soldered together, their backs forming one continuous shield—the carapace.

Unity of plan, diversity in execution, is the lesson taught by the study of the rings of the body, and the same instruction is given still more emphatically by the appendages. If I examine the outermost jaw I find it consists of three distinct portions, an inner, a middle, and an outer, mounted

upon a common stem; and if I compare this jaw
with the legs behind it, or the jaws in front of it,
I find it quite easy to see, that, in the legs, it is
the part of the appendage which corresponds with
the inner division, which becomes modified into
what we know familiarly as the "leg," while the
middle division disappears, and the outer division
is hidden under the carapace. Nor is it more
difficult to discern that, in the appendages of the
tail, the middle division appears again and the
outer vanishes; while, on the other hand, in the
foremost jaw, the so-called mandible, the inner
division only is left; and, in the same way, the
parts of the feelers and of the eye-stalks can be
identified with those of the legs and jaws.

But whither does all this tend? To the very
remarkable conclusion that a unity of plan, of the
same kind as that discoverable in the tail or
abdomen of the lobster, pervades the whole organisation of its skeleton, so that I can return to the
diagram representing any one of the rings of the
tail, which I drew upon the board, and by adding a
third division to each appendage, I can use it as a
sort of scheme or plan of any ring of the body. I
can give names to all the parts of that figure, and
then if I take any segment of the body of the
lobster, I can point out to you exactly, what modification the general plan has undergone in that
particular segment; what part has remained
movable, and what has become fixed to another;

what has been excessively developed and metamorphosed and what has been suppressed.

But I imagine I hear the question, How is all
this to be tested? No doubt it is a pretty and
ingenious way of looking at the structure of any
animal; but is it anything more? Does Nature
acknowledge, in any deeper way, this unity of plan
we seem to trace?

The objection suggested by these questions is a
very valid and important one, and morphology was
in an unsound state so long as it rested upon the
mere perception of the analogies which obtain
between fully formed parts. The unchecked ingenuity of speculative anatomists proved itself
fully competent to spin any number of contradictory hypotheses out of the same facts, and endless
morphological dreams threatened to supplant
scientific theory.

Happily, however, there is a criterion of morphological truth, and a sure test of all homologies.
Our lobster has not always been what we see it;
it was once an egg, a semifluid mass of yolk, not so
big as a pin's head, contained in a transparent
membrane, and exhibiting not the least trace of
any one of those organs, the multiplicity and
complexity of which, in the adult, are so surprising.
After a time, a delicate patch of cellular membrane
appeared upon one face of this yolk, and that
patch was the foundation of the whole creature,
the clay out of which it would be moulded.

Gradually investing the yolk, it became subdivided by transverse constrictions into segments, the forerunners of the rings of the body. Upon the ventral surface of each of the rings thus sketched out, a pair of bud-like prominences made their appearance—the rudiments of the appendages of the ring. At first, all the appendages were alike, but, as they grew, most of them became distinguished into a stem and two terminal divisions, to which, in the middle part of the body, was added a third outer division; and it was only at a later period, that by the modification, or absorption, of certain of these primitive constituents, the limbs acquired their perfect form.

Thus the study of development proves that the doctrine of unity of plan is not merely a fancy, that it is not merely one way of looking at the matter, but that it is the expression of deep-seated natural facts. The legs and jaws of the lobster may not merely be regarded as modifications of a common type,—in fact and in nature they are so, —the leg and the jaw of the young animal being, at first, indistinguishable.

These are wonderful truths, the more so because the zoologist finds them to be of universal application. The investigation of a polype, of a snail, of a fish, of a horse, or of a man, would have led us, though by a less easy path, perhaps, to exactly the same point. Unity of plan everywhere lies hidden under the mask of diversity of structure—the complex is everywhere evolved out of the simple. Every animal has at first the form of an egg, and every animal and every organic part, in reaching its adult state, passes through conditions common to other animals and other adult parts; and this leads me to another point. I have hitherto spoken as if the lobster were alone in the world, but, as I need hardly remind you, there are myriads of other animal organisms. Of these, some, such as men, horses, birds, fishes, snails, slugs, oysters, corals, and sponges, are not in the least like the lobster. But other animals, though they may differ a good deal from the lobster, are yet either very like it, or are like something that is like it. The cray fish, the rock lobster, and the prawn, and the shrimp, for example, however different, are yet so like lobsters, that a child would group them as of the lobster kind, in contradistinction to snails and slugs; and these last again would form a kind by themselves, in contradistinction to cows, horses, and sheep, the cattle kind.

But this spontaneous grouping into "kinds" is the first essay of the human mind at classification, or the calling by a common name of those things that are alike, and the arranging them in such a manner as best to suggest the sum of their likenesses and unlikenesses to other things.

Those kinds which include no other subdivisions than the sexes, or various breeds, are called, in technical language, species. The English lobster

is a species, our cray fish is another, our prawn is
another. In other countries, however, there are
lobsters, cray fish, and prawns, very like ours, and
yet presenting sufficient differences to deserve distinction. Naturalists, therefore, express this resemblance and this diversity by grouping them as
distinct species of the same "genus." But the
lobster and the cray fish, though belonging to distinct genera, have many features in common, and
hence are grouped together in an assemblage which
is called a family. More distant resemblances
connect the lobster with the prawn and the crab,
which are expressed by putting all these into the
same order. Again, more remote, but still very
definite, resemblances unite the lobster with the
woodlouse, the king crab, the water flea, and the
barnacle, and separate them from all other animals ;
whence they collectively constitute the larger
group, or class, *Crustacea*. But the *Crustacea*
exhibit many peculiar features in common with
insects, spiders, and centipedes, so that these are
grouped into the still larger assemblage or " province " *Articulata ;* and, finally, the relations
which these have to worms and other lower
animals, are expressed by combining the whole vast
aggregate into the sub-kingdom of *Annulosa*.

If I had worked my way from a sponge instead
of a lobster, I should have found it associated, by
like ties, with a great number of other animals
into the sub-kingdom *Protozoa ;* if I had selected
a fresh-water polype or a coral, the members of

what naturalists term the sub-kingdom *Cœlenterata*, would have grouped themselves around my type; had a snail been chosen, the inhabitants of all univalve and bivalve, land and water, shells, the lamp shells, the squids, and the sea-mat would have gradually linked themselves on to it as members of the same sub-kingdom of *Mollusca*; and finally, starting from man, I should have been compelled to admit first, the ape, the rat, the horse, the dog, into the same class; and then the bird, the crocodile, the turtle, the frog, and the fish, into the same sub-kingdom of *Vertebrata*.

And if I had followed out all these various lines of classification fully, I should discover in the end that there was no animal, either recent or fossil, which did not at once fall into one or other of these sub-kingdoms. In other words, every animal is organised upon one or other of the five, or more, plans, the existence of which renders our classification possible. And so definitely and precisely marked is the structure of each animal, that, in the present state of our knowledge, there is not the least evidence to prove that a form, in the slightest degree transitional between any of the two groups *Vertebrata*, *Annulosa*, *Mollusca*, and *Cœlenterata*, either exists, or has existed, during that period of the earth's history which is recorded by the geologist.[1] Nevertheless, you must not for a moment suppose, because no such

[1 The different grouping necessitated by later knowledge does not affect the principle of the argument.—1894.]

transitional forms are known, that the members of the sub-kingdoms are disconnected from, or independent of, one another. On the contrary, in their earliest condition they are all similar, and the primordial germs of a man, a dog, a bird, a fish, a beetle, a snail, and a polype are, in no essential structural respects, distinguishable.

In this broad sense, it may with truth be said, that all living animals, and all those dead faunæ which geology reveals, are bound together by an all-pervading unity of organisation, of the same character, though not equal in degree, to that which enables us to discern one and the same plan amidst the twenty different segments of a lobster's body. Truly it has been said, that to a clear eye the smallest fact is a window through which the Infinite may be seen.

Turning from these purely morphological considerations, let us now examine into the manner in which the attentive study of the lobster impels us into other lines of research.

Lobsters are found in all the European seas; but on the opposite shores of the Atlantic and in the seas of the southern hemisphere they do not exist. They are, however, represented in these regions by very closely allied, but distinct forms— the *Homarus Americanus* and the *Homarus Capensis*: so that we may say that the European has one species of *Homarus*; the American, another; the African, another; and thus the

remarkable facts of geographical distribution begin
to dawn upon us.

Again, if we examine the contents of the earth's
crust, we shall find in the latter of those deposits,
which have served as the great burying grounds of
past ages, numberless lobster-like animals, but
none so similar to our living lobster as to make
zoologists sure that they belonged even to the
same genus. If we go still further back in time,
we discover, in the oldest rocks of all, the remains
of animals, constructed on the same general plan
as the lobster, and belonging to the same great
group of *Crustacea;* but for the most part
totally different from the lobster, and indeed from
any other living form of crustacean; and thus we
gain a notion of that successive change of the
animal population of the globe, in past ages,
which is the most striking fact revealed by
geology.

Consider, now, where our inquiries have led us.
We studied our type morphologically, when we
determined its anatomy and its development, and
when comparing it, in these respects, with other
animals, we made out its place in a system of
classification. If we were to examine every
animal in a similar manner, we should establish a
complete body of zoological morphology.

Again, we investigated the distribution of our
type in space and in time, and, if the like had
been done with every animal, the sciences of geo-

graphical and geological distribution would have attained their limit.

But you will observe one remarkable circumstance, that, up to this point, the question of the life of these organisms has not come under consideration. Morphology and distribution might be studied almost as well, if animals and plants were a peculiar kind of crystals, and possessed none of those functions which distinguish living beings so remarkably. But the facts of morphology and distribution have to be accounted for, and the science, the aim of which it is to account for them, is Physiology.

Let us return to our lobster once more. If we watched the creature in its native element, we should see it climbing actively the submerged rocks, among which it delights to live, by means of its strong legs; or swimming by powerful strokes of its great tail, the appendages of the sixth joint of which are spread out into a broad fan-like propeller: seize it, and it will show you that its great claws are no mean weapons of offence; suspend a piece of carrion among its haunts, and it will greedily devour it, tearing and crushing the flesh by means of its multitudinous jaws.

Suppose that we had known nothing of the lobster but as an inert mass, an organic crystal, if I may use the phrase, and that we could suddenly see it exerting all these powers, what wonderful new ideas and new questions would arise in our

minds! The great new question would be, "How
does all this take place?" the chief new idea would
be, the idea of adaptation to purpose,—the notion,
that the constituents of animal bodies are not
mere unconnected parts, but organs working
together to an end. Let us consider the tail of
the lobster again from this point of view.
Morphology has taught us that it is a series
of segments composed of homologous parts,
which undergo various modifications—beneath
and through which a common plan of formation
is discernible. But if I look at the same part
physiologically, I see that it is a most beautifully
constructed organ of locomotion, by means of
which the animal can swiftly propel itself either
backwards or forwards.

But how is this remarkable propulsive machine
made to perform its functions? If I were suddenly to kill one of these animals and to take out
all the soft parts, I should find the shell to be perfectly inert, to have no more power of moving
itself than is possessed by the machinery of a mill
when disconnected from its steam-engine or water-
wheel. But if I were to open it, and take out the
viscera only, leaving the white flesh, I should perceive that the lobster could bend and extend its
tail as well as before. If I were to cut off the
tail, I should cease to find any spontaneous motion
in it; but on pinching any portion of the flesh,
I should observe that it underwent a very curious

change—each fibre becoming shorter and thicker. By this act of contraction, as it is termed, the parts to which the ends of the fibre are attached are, of course, approximated ; and according to the relations of their points of attachment to the centres of motions of the different rings, the bending or the extension of the tail results. Close observation of the newly-opened lobster would soon show that all its movements are due to the same cause—the shortening and thickening of these fleshy fibres, which are technically called muscles.

Here, then, is a capital fact. The movements of the lobster are due to muscular contractility. But why does a muscle contract at one time and not at another ? Why does one whole group of muscles contract when the lobster wishes to extend his tail, and another group when he desires to bend it ? What is it originates, directs, and controls the motive power ?

Experiment, the great instrument for the ascertainment of truth in physical science, answers this question for us. In the head of the lobster there lies a small mass of that peculiar tissue which is known as nervous substance. Cords of similar matter connect this brain of the lobster, directly or indirectly, with the muscles. Now, if these communicating cords are cut, the brain remaining entire, the power of exerting what we call voluntary motion in the parts below the section is destroyed ; and, on the other hand, if, the cords remaining entire, the brain mass be destroyed, the same voluntary mobility is equally lost. Whence the inevitable conclusion is, that the power of originating these motions resides in the brain and is propagated along the nervous cords.

In the higher animals the phenomena which attend this transmission have been investigated, and the exertion of the peculiar energy which resides in the nerves has been found to be accompanied by a disturbance of the electrical state of their molecules.

If we could exactly estimate the signification of this disturbance ; if we could obtain the value of a given exertion of nerve force by determining the quantity of electricity, or of heat, of which it is the equivalent ; if we could ascertain upon what arrangement, or other condition of the molecules of matter, the manifestation of the nervous and muscular energies depends (and doubtless science will some day or other ascertain these points), physiologists would have attained their ultimate goal in this direction ; they would have determined the relation of the motive force of animals to the other forms of force found in nature ; and if the same process had been successfully performed for all the operations which are carried on in, and by, the animal frame, physiology would be perfect, and the facts of morphology and distribution would be deducible from the laws which physiologists had established, combined with those determining the condition of the surrounding universe.

There is not a fragment of the organism of this humble animal whose study would not lead us into regions of thought as large as those which I have briefly opened up to you; but what I have been saying, I trust, has not only enabled you to form a conception of the scope and purport of zoology, but has given you an imperfect example of the manner in which, in my opinion, that science, or indeed any physical science, may be best taught. The great matter is, to make teaching real and practical, by fixing the attention of the student on particular facts; but at the same time it should be rendered broad and comprehensive, by constant reference to the generalisations of which all particular facts are illustrations. The lobster has served as a type of the whole animal kingdom, and its anatomy and physiology have illustrated for us some of the greatest truths of biology. The student who has once seen for himself the facts which I have described, has had their relations explained to him, and has clearly comprehended them, has, so far, a knowledge of zoology, which is real and genuine, however limited it may be, and which is worth more than all the mere reading knowledge of the science he could ever acquire. His zoological information is, so far, knowledge and not mere hearsay.

And if it were my business to fit you for the certificate in zoological science granted by this department, I should pursue a course precisely similar in principle to that which I have taken to-night. I should select a fresh-water sponge, a fresh-water polype or a *Cyanœa*, a fresh-water mussel, a lobster, a fowl, as types of the five primary divisions of the animal kingdom. I should explain their structure very fully, and show how each illustrated the great principles of zoology. Having gone very carefully and fully over this ground, I should feel that you had a safe foundation, and I should then take you in the same way, but less minutely, over similarly selected illustrative types of the classes; and then I should direct your attention to the special forms enumerated under the head of types, in this syllabus, and to the other facts there mentioned.

That would, speaking generally, be my plan. But I have undertaken to explain to you the best mode of acquiring and communicating a knowledge of zoology, and you may therefore fairly ask me for a more detailed and precise account of the manner in which I should propose to furnish you with the information I refer to.

My own impression is, that the best model for all kinds of training in physical science is that afforded by the method of teaching anatomy, in use in the medical schools. This method consists of three elements—lectures, demonstrations, and examinations.

The object of lectures is, in the first place, to awaken the attention and excite the enthusiasm of the student; and this, I am sure, may be effected to a far greater extent by the oral discourse and by the personal influence of a respected teacher than in any other way. Secondly, lectures have the double use of guiding the student to the salient points of a subject, and at the same time forcing him to attend to the whole of it, and not merely to that part which takes his fancy. And lastly, lectures afford the student the opportunity of seeking explanations of those difficulties which will, and indeed ought to, arise in the course of his studies.

What books shall I read? is a question constantly put by the student to the teacher. My reply usually is, " None: write your notes out carefully and fully; strive to understand them thoroughly; come to me for the explanation of anything you cannot understand; and I would rather you did not distract your mind by reading." A properly composed course of lectures ought to contain fully as much matter as a student can assimilate in the time occupied by its delivery; and the teacher should always recollect that his business is to feed, and not to cram the intellect.

Indeed, I believe that a student who gains from a course of lectures the simple habit of concentrating his attention upon a definitely limited series of facts, until they are thoroughly mastered, has made a step of immeasurable importance.

But, however good lectures may be, and however extensive the course of reading by which they are followed up, they are but accessories to the great instrument of scientific teaching—demonstration. If I insist unweariedly, nay fanatically, upon the importance of physical science as an educational agent, it is because the study of any branch of science, if properly conducted, appears to me to fill up a void left by all other means of education. I have the greatest respect and love for literature; nothing would grieve me more than to see literary training other than a very prominent branch of education: indeed, I wish that real literary discipline were far more attended to than it is; but I cannot shut my eyes to the fact, that there is a vast difference between men who have had a purely literary, and those who have had a sound scientific, training.

Seeking for the cause of this difference, I imagine I can find it in the fact that, in the world of letters, learning and knowledge are one, and books are the source of both; whereas in science, as in life, learning and knowledge are distinct, and the study of things, and not of books, is the source of the latter.

All that literature has to bestow may be obtained by reading and by practical exercise in writing and in speaking; but I do not exaggerate when I say, that none of the best gifts of science are to be won by these means. On the contrary, the great benefit which a scientific education bestows, whether as training or as knowledge, is dependent upon the extent to which the mind of the student is brought into immediate contact with facts— upon the degree to which he learns the habit of appealing directly to Nature, and of acquiring through his senses concrete images of those properties of things, which are, and always will be, but approximatively expressed in human language. Our way of looking at Nature, and of speaking about her, varies from year to year; but a fact once seen, a relation of cause and effect, once demonstratively apprehended, are possessions which neither change nor pass away, but, on the contrary, form fixed centres, about which other truths aggregate by natural affinity.

Therefore, the great business of the scientific teacher is, to imprint the fundamental, irrefragable facts of his science, not only by words upon the mind, but by sensible impressions upon the eye, and ear, and touch of the student, in so complete a manner, that every term used, or law enunciated, should afterwards call up vivid images of the

particular structural, or other, facts which furnished
the demonstration of the law, or the illustration
of the term.

Now this important operation can only be
achieved by constant demonstration, which may
take place to a certain imperfect extent during a
lecture, but which ought also to be carried on
independently, and which should be addressed to
each individual student, the teacher endeavouring,
not so much to show a thing to the learner, as to
make him see it for himself.

I am well aware that there are great practical
difficulties in the way of effectual zoological
demonstrations. The dissection of animals is not
altogether pleasant, and requires much time; nor
is it easy to secure an adequate supply of the
needful specimens. The botanist has here a
great advantage; his specimens are easily obtained, are clean and wholesome, and can be
dissected in a private house as well as anywhere
else; and hence, I believe, the fact, that botany
is so much more readily and better taught than
its sister science. But, be it difficult or be it
easy, if zoological science is to be properly studied,
demonstration, and, consequently, dissection, must
be had. Without it, no man can have a really
sound knowledge of animal organisation.

A good deal may be done, however, without
actual dissection on the student's part, by demonstration upon specimens and preparations; and in

all probability it would not be very difficult, were
the demand sufficient, to organise collections of
such objects, sufficient for all the purposes of
elementary teaching, at a comparatively cheap
rate.   Even without these, much might be
effected, if the zoological collections, which are
open to the public, were arranged according to
what has been termed the "typical principle";
that is to say, if the specimens exposed to public
view were so selected that the public could learn
something from them, instead of being, as at
present, merely confused by their multiplicity.
For example, the grand ornithological gallery at
the British Museum contains between two and
three thousand species of birds, and sometimes
five or six specimens of a species.   They are
very pretty to look at, and some of the cases are,
indeed, splendid; but I will undertake to say,
that no man but a professed ornithologist has
ever gathered much information from the collection.   Certainly, no one of the tens of thousands
of the general public who have walked through
that gallery ever knew more about the essential
peculiarities of birds when he left the gallery
than when he entered it.   But if, somewhere in
that vast hall, there were a few preparations,
exemplifying the leading structural peculiarities
and the mode of development of a common fowl;
if the types of the genera, the leading modifications in the skeleton, in the plumage at various

ages, in the mode of nidification, and the like, among birds, were displayed; and if the other specimens were put away in a place where the men of science, to whom they are alone useful, could have free access to them, I can conceive that this collection might become a great instrument of scientific education.

The last implement of the teacher to which I have adverted is examination—a means of education now so thoroughly understood that I need hardly enlarge upon it. I hold that both written and oral examinations are indispensable, and, by requiring the description of specimens, they may be made to supplement demonstration.

Such is the fullest reply the time at my disposal will allow me to give to the question—how may a knowledge of zoology be best acquired and communicated?

But there is a previous question which may be moved, and which, in fact, I know many are inclined to move. It is the question, why should teachers be encouraged to acquire a knowledge of this, or any other branch of physical science? What is the use, it is said, of attempting to make physical science a branch of primary education? Is it not probable that teachers, in pursuing such studies, will be led astray from the acquirement of more important but less attractive knowledge? And, even if they can learn something of science without prejudice to their useful-

ness, what is the good of their attempting to
instil that knowledge into boys whose real business is the acquisition of reading, writing, and
arithmetic ?

These questions are, and will be, very commonly
asked, for they arise from that profound ignorance
of the value and true position of physical science,
which infests the minds of the most highly educated and intelligent classes of the community.
But if I did not feel well assured that they are
capable of being easily and satisfactorily answered ;
that they have been answered over and over again ;
and that the time will come when men of liberal
education will blush to raise such questions—I
should be ashamed of my position here to-night.
Without doubt, it is your great and very important
function to carry out elementary education ; without question, anything that should interfere with
the faithful fulfilment of that duty on your part
would be a great evil ; and if I thought that your
acquirement of the elements of physical science,
and your communication of those elements to your
pupils, involved any sort of interference with your
proper duties, I should be the first person to pro-
test against your being encouraged to do anything
of the kind.

But is it true that the acquisition of such a
knowledge of science as is proposed, and the communication of that knowledge, are calculated to
weaken your usefulness ?   Or may I not rather

ask, is it possible for you to discharge your functions properly without these aids?

What is the purpose of primary intellectual education? I apprehend that its first object is to train the young in the use of those tools wherewith men extract knowledge from the ever-shifting succession of phenomena which pass before their eyes; and that its second object is to inform them of the fundamental laws which have been found by experience to govern the course of things, so that they may not be turned out into the world naked, defenceless, and a prey to the events they might control.

A boy is taught to read his own and other languages, in order that he may have access to infinitely wider stores of knowledge than could ever be opened to him by oral intercourse with his fellow men; he learns to write, that his means of communication with the rest of mankind may be indefinitely enlarged, and that he may record and store up the knowledge he acquires. He is taught elementary mathematics, that he may understand all those relations of number and form, upon which the transactions of men, associated in complicated societies, are built, and that he may have some practice in deductive reasoning.

All these operations of reading, writing, and ciphering, are intellectual tools, whose use should, before all things, be learned, and learned thoroughly; so that the youth may be enabled to make his life that which it ought to be, a continual progress in learning and in wisdom.

But, in addition, primary education endeavours to fit a boy out with a certain equipment of positive knowledge. He is taught the great laws of morality; the religion of his sect; so much history and geography as will tell him where the great countries of the world are, what they are, and how they have become what they are.

Without doubt all these are most fitting and excellent things to teach a boy; I should be very sorry to omit any of them from any scheme of primary intellectual education. The system is excellent, so far as it goes.

But if I regard it closely, a curious reflection arises. I suppose that, fifteen hundred years ago, the child of any well-to-do Roman citizen was taught just these same things; reading and writing in his own, and, perhaps, the Greek tongue; the elements of mathematics; and the religion, morality, history, and geography current in his time. Furthermore, I do not think I err in affirming, that, if such a Christian Roman boy, who had finished his education, could be transplanted into one of our public schools, and pass through its course of instruction, he would not meet with a single unfamiliar line of thought; amidst all the new facts he would have to learn, not one would suggest a different mode of regarding the universe from that current in his own time.

And yet surely there is some great difference between the civilisation of the fourth century and that of the nineteenth, and still more between the intellectual habits and tone of thought of that day and this?

And what has made this difference? I answer fearlessly—The prodigious development of physical science within the last two centuries.

Modern civilisation rests upon physical science; take away her gifts to our own country, and our position among the leading nations of the world is gone to-morrow; for it is physical science only that makes intelligence and moral energy stronger than brute force.

The whole of modern thought is steeped in science; it has made its way into the works of our best poets, and even the mere man of letters, who affects to ignore and despise science, is unconsciously impregnated with her spirit, and indebted for his best products to her methods. I believe that the greatest intellectual revolution mankind has yet seen is now slowly taking place by her agency. She is teaching the world that the ultimate court of appeal is observation and experiment, and not authority; she is teaching it to estimate the value of evidence; she is creating a firm and living faith in the existence of immutable moral and physical laws, perfect obedience to which is the highest possible aim of an intelligent being.

But of all this your old stereotyped system of education takes no note. Physical science, its methods, its problems, and its difficulties, will meet the poorest boy at every turn, and yet we educate him in such a manner that he shall enter the world as ignorant of the existence of the methods and facts of science as the day he was born. The modern world is full of artillery; and we turn out our children to do battle in it, equipped with the shield and sword of an ancient gladiator.

Posterity will cry shame on us if we do not remedy this deplorable state of things. Nay, if we live twenty years longer, our own consciences will cry shame on us.

It is my firm conviction that the only way to remedy it is to make the elements of physical science an integral part of primary education. I have endeavoured to show you how that may be done for that branch of science which it is my business to pursue; and I can but add, that I should look upon the day when every schoolmaster throughout this land was a centre of genuine, however rudimentary, scientific knowledge, as an epoch in the history of the country.

But let me entreat you to remember my last words. Addressing myself to you, as teachers, I would say, mere book learning in physical science is a sham and a delusion—what you teach, unless you wish to be impostors, that you must first know; and real knowledge in science means personal acquaintance with the facts, be they few or many.[1]

[1] It has been suggested to me that these words may be taken to imply a discouragement on my part of any sort of scientific instruction which does not give an acquaintance with the facts at first hand. But this is not my meaning. The ideal of scientific teaching is, no doubt, a system by which the scholar sees every fact for himself, and the teacher supplies only the explanations. Circumstances, however, do not often allow of the attainment of that ideal, and we must put up with the next best system—one in which the scholar takes a good deal on trust from a teacher, who, knowing the facts by his own knowledge, can describe them with so much vividness as to enable his audience to form competent ideas concerning them. The system which I repudiate is that which allows teachers who have not come into direct contact with the leading facts of a science to pass their second-hand information on. The scientific virus, like vaccine lymph, if passed through too long a succession of organisms, will lose all its effect in protecting the young against the intellectual epidemics to which they are exposed.

[The remarks on p. 222 applied to the Natural History Collection of the British Museum in 1861. The visitor to the Natural History Museum in 1894 need go no further than the Great Hall to see the realisation of my hopes by the present Director.]

VIII

BIOGENESIS AND ABIOGENESIS

(THE PRESIDENTIAL ADDRESS TO THE BRITISH
ASSOCIATION FOR THE ADVANCEMENT OF SCIENCE
FOR 1870)

IT has long been the custom for the newly
installed President of the British Association for
the Advancement of Science to take advantage of
the elevation of the position in which the suffrages
of his colleagues had, for the time, placed him, and,
casting his eyes around the horizon of the scientific
world, to report to them what could be seen from
his watch-tower; in what directions the multitudinous divisions of the noble army of the improvers
of natural knowledge were marching; what
important strongholds of the great enemy of us
all, ignorance, had been recently captured; and,
also, with due impartiality, to mark where the
advanced posts of science had been driven in, or a
long-continued siege had made no progress.

I propose to endeavour to follow this ancient precedent, in a manner suited to the limitations of my knowledge and of my capacity. I shall not presume to attempt a panoramic survey of the world of science, nor even to give a sketch of what is doing in the one great province of biology, with some portions of which my ordinary occupations render me familiar. But I shall endeavour to put before you the history of the rise and progress of a single biological doctrine; and I shall try to give some notion of the fruits, both intellectual and practical, which we owe, directly or indirectly, to the working out, by seven generations of patient and laborious investigators, of the thought which arose, more than two centuries ago, in the mind of a sagacious and observant Italian naturalist.

It is a matter of everyday experience that it is difficult to prevent many articles of food from becoming covered with mould; that fruit, sound enough to all appearance, often contains grubs at the core; that meat, left to itself in the air, is apt to putrefy and swarm with maggots. Even ordinary water, if allowed to stand in an open vessel, sooner or later becomes turbid and full of living matter.

The philosophers of antiquity, interrogated as to the cause of these phenomena, were provided with a ready and a plausible answer. It did not enter their minds even to doubt that these low forms of life were generated in the matters in which they made their appearance. Lucretius, who had drunk deeper of the scientific spirit than any poet of ancient or modern times except Goethe, intends to speak as a philosopher, rather than as a poet, when he writes that "with good reason the earth has gotten the name of mother, since all things are produced out of the earth. And many living creatures, even now, spring out of the earth, taking form by the rains and the heat of the sun." [1] The axiom of ancient science, "that the corruption of one thing is the birth of another," had its popular embodiment in the notion that a seed dies before the young plant springs from it; a belief so wide-spread and so fixed, that Saint Paul appeals to it in one of the most splendid outbursts of his fervid eloquence :—

"Thou fool, that which thou sowest is not quickened, except it die." [2]

The proposition that life may, and does, proceed from that which has no life, then, was held alike by the philosophers, the poets, and the people, of

[1] It is thus that Mr. Munro renders

"Linquitur, ut merito maternum nomen adepta
Terra sit, e terra quoniam sunt cuncta creata.
Multaque nunc etiam exsistant animalia terris
Imbribus et calido solis concreta vapore."
De Rerum Natura, lib. v. 793—796.

But would not the meaning of the last line be better rendered "Developed in rain-water and in the warm vapours raised by the sun"?   [2] 1 Corinthians xv. 36.

the most enlightened nations, eighteen hundred years ago; and it remained the accepted doctrine of learned and unlearned Europe, through the Middle Ages, down even to the seventeenth century.

It is commonly counted among the many merits of our great countryman, Harvey, that he was the first to declare the opposition of fact to venerable authority in this, as in other matters; but I can discover no justification for this widespread notion. After careful search through the "Exercitationes de Generatione," the most that appears clear to me is, that Harvey believed all animals and plants to spring from what he terms a "*prim-ordium vegetale,*" a phrase which may nowadays be rendered "a vegetative germ"; and this, he says, is "*oviforme,*" or "egg-like"; not, he is careful to add, that it necessarily has the shape of an egg, but because it has the constitution and nature of one. That this "*primordium oviforme*" must needs, in all cases, proceed from a living parent is nowhere expressly maintained by Harvey, though such an opinion may be thought to be implied in one or two passages; while, on the other hand, he does, more than once, use language which is consistent only with a full belief in spontaneous or equivocal generation.[1]   In fact, the main concern of Harvey's

---

[1] See the following passage in Exercitatio I. :—"Item *sponte nascentia* dicuntur; non quod ex *putredine* oriunda sint, sed quod casu, naturæ sponte, et æquivocâ (ut aiunt) generatione, a

wonderful little treatise is not with generation, in the physiological sense, at all, but with development; and his great object is the establishment of the doctrine of epigenesis.

The first distinct enunciation of the hypothesis that all living matter has sprung from pre-existing living matter, came from a contemporary, though a junior, of Harvey, a native of that country, fertile in men great in all departments of human activity, which was to intellectual Europe, in the sixteenth and seventeenth centuries, what Germany is in the nineteenth.   It was in Italy, and from Italian teachers, that Harvey received the most important part of his scientific education.   And it was a student trained in the same schools, Francesco Redi—a man of the widest knowledge and most versatile abilities, distinguished alike as scholar, poet, physician, and naturalist—who, just two hundred and two years ago, published his " Esperienze intorno alla Generazione degl' Insetti," and gave to the world the idea, the growth of which it is my purpose to trace.   Redi's book went through five editions in twenty years; and the extreme

parentibus sui dissimilibus proveniant." Again, in *De Uteri Membranis :*—" In cunctorum viventium generatione (sicut diximus) hoc solenne est, ut ortum ducunt a *primordio* aliquo, quod tum materiam tum efficiendi potestatem in se habet : sitque adeo id, ex quo et a quo quicquid nascitur, ortum suum ducat. Tale primordium in animalibus (*sive ab aliis generantibus proveniant, sive sponte, aut ex putredine nascentur*) est humor in tunicâ aliquâ aut putami ne conclusus." Compare also what Redi has to say respecting Harvey's opinions, *Esperienze,* p. 11.

simplicity of his experiments, and the clearness of his arguments, gained for his views, and for their consequences, almost universal acceptance.

Redi did not trouble himself much with speculative considerations, but attacked particular cases of what was supposed to be "spontaneous generation" experimentally. Here are dead animals, or pieces of meat, says he ; I expose them to the air in hot weather, and in a few days they swarm with maggots. You tell me that these are generated in the dead flesh; but if I put similar bodies, while quite fresh, into a jar, and tie some fine gauze over the top of the jar, not a maggot makes its appearance, while the dead substances, nevertheless, putrefy just in the same way as before. It is obvious, therefore, that the maggots are not generated by the corruption of the meat ; and that the cause of their formation must be a something which is kept away by gauze. But gauze will not keep away aëriform bodies, or fluids. This something must, therefore, exist in the form of solid particles too big to get through the gauze. Nor is one long left in doubt what these solid particles are ; for the blow-flies, attracted by the odour of the meat, swarm round the vessel, and, urged by a powerful but in this case misleading instinct, lay eggs out of which maggots are immediately hatched, upon the gauze. The conclusion, therefore, is unavoidable; the maggots are not generated by the meat, but the eggs which give rise to them are brought through the air by the flies.

These experiments seem almost childishly simple, and one wonders how it was that no one ever thought of them before. Simple as they are, however, they are worthy of the most careful study, for every piece of experimental work since done, in regard to this subject, has been shaped upon the model furnished by the Italian philosopher. As the results of his experiments were the same, however varied the nature of the materials he used, it is not wonderful that there arose in Redi's mind a presumption, that, in all such cases of the seeming production of life from dead matter, the real explanation was the introduction of living germs from without into that dead matter.[1]

---

[1] "Pure contentandomi sempre in questa ed in ciascuna altro cosa, da ciascuno più savio, là dove io difettuosamente parlassi, esser corretto ; non tacero, che per molte osservazioni molti volti da me fatte, mi sento inclinato a credere che la terra, da quelle prime piante, e da quei primi animali in poi, che ella nei primi giorni del mondo produsse per comandemento del sovrano ed onnipotente Fattore, non abbia mai più prodotto da se medesima nè erba nè albero, nè animale alcuno perfetto o imperfetto che ei se fosse ; e che tutto quello, che ne' tempi trapassati è nato e che ora nascere in lei, o da lei veggiamo, venga tutto dalla semenza reale e vera delle piante, e degli animali stessi, i quali col mezzo del proprio seme la loro spezie conservano. E se bene tutto giorno scorghiamo da' cadaveri degli animali, e da tutte quante le maniere dell' erbe, e de' fiori, e dei frutti imputriditi, e corrotti nascere vermi infiniti—
   'Nonne vides quæcunque mora, fluidoque calore
   Corpora tabescunt in parva animalia verti '—
Io mi sento, dico, inclinato, a credere che tutti quei vermi si

And thus the hypothesis that living matter always arises by the agency of pre-existing living matter, took definite shape ; and had, henceforward, a right to be considered and a claim to be refuted, in each particular case, before the production of living matter in any other way could be admitted by careful reasoners. It will be necessary for me to refer to this hypothesis so frequently, that, to save circumlocution, I shall call it the hypothesis of *Biogenesis ;* and I shall term the contrary doctrine —that living matter may be produced by not living matter—the hypothesis of *Abiogenesis.*

In the seventeenth century, as I have said, the latter was the dominant view, sanctioned alike by antiquity and by authority; and it is interesting to observe that Redi did not escape the customary tax upon a discoverer of having to defend himself against the charge of impugning the authority of the Scriptures ; [1] for his adversaries declared that

generino dal seme paterno ; e che le carni, e l' erbe, e l' altre cose tutte putrefatte, o putrefattibili non facciano altra parte, nè abbiano altro ufizio nella generazione degl' insetti, se non d'apprestare un luogo o un nido proporzionato, in cui dagli animali nel tempo della figliatura sieno portati, e partoriti i vermi, o l' uova o l' altre semenze dei vermi, i quali tosto che nati sono, trovano in esso nido un sufficiente alimento abilissimo per nutricarsi : e se in quello non son portate dalle madri queste suddette semenze, niente mai, e replicatamente niente, vi s' ingegneri e nasca."—REDI, *Esperienze,* pp. 14–16.

[1] " Molti, e molti altri ancora vi potrei annoverare, se non fossi chiamato a rispondere alle rampogne di alcuni, che bruscamente mi rammentano ciò, che si legge nel capitolo quattordicesimo del sacrosanto Libro de' giudici. . . ."—REDI, *loc. cit.* p. 45.

the generation of bees from the carcase of a dead
lion is affirmed, in the Book of Judges, to have
been the origin of the famous riddle with which
Samson perplexed the Philistines :—

> " Out of the eater came forth meat,
> And out of the strong came forth sweetness.

Against all odds, however, Redi, strong with
the strength of demonstrable fact, did splendid
battle for Biogenesis; but it is remarkable that
he held the doctrine in a sense which, if he had
lived in these times, would have infallibly caused
him to be classed among the defenders of " spontaneous generation."   " Omne vivum ex vivo,"
" no life without antecedent life," aphoristically
sums up Redi's doctrine; but he went no further.
It is most remarkable evidence of the philosophic
caution and impartiality of his mind, that although
he had speculatively anticipated the manner in
which grubs really are deposited in fruits and in
the galls of plants, he deliberately admits that the
evidence is insufficient to bear him out; and he
therefore prefers the supposition that they are
generated by a modification of the living substance
of the plants themselves.   Indeed, he regards
these vegetable growths as organs, by means of
which the plant gives rise to an animal, and looks
upon this production of specific animals as the
final cause of the galls and of, at any rate, some
fruits.   And he proposes to explain the occurrence

of parasites within the animal body in the same way.[1]

[1] The passage (*Esperienze*, p. 129) is worth quoting in full :—
    " Se dovessi palesarvi il mio sentimento crederei che i frutti, i legumi, gli alberi e le foglie, in due maniere inverminassero. Una, perchè venendo i bachi per di fuora, e cercando l' alimento, col rodere ci aprono la strada, ed arrivano alla più interna midolla de' frutti e de' legni.   L'altra maniera si è, che io per me stimerei, che non fosse gran fatto disdicevole il credere, che quell' anima o quella virtù, la quale genera i fiori ed i frutti nelle piante viventi, sia quella stessa che generi ancora i bachi di esse piante.   E chi sà forse, che molti frutti degli alberi non sieno prodotti, non per un fine primario e principale, ma bensi per un uffizio secondario e servile, destinato alla generazione di que' vermi, servendo a loro in vece di matrice, in cui dimorino un prefisso e determinato tempo ; il quale arrivato escan fuora a godere il sole.
    " Io m' immagino, che questo mio pensiero non vi parrà totalmente un paradosso ; mentre farete riflessione a quelle tante sorte di galle, di gallozzole, di coccole, di ricci, di calici, di cornetti ed i lappole, che son produtte dalle quercel, dalle farnie, da' cerri, da' sugheri, da' lecci e da altri simili alberi da ghianda ; imperciocchè in quelle gallozzole, e particolarmente nelle più grosse, che si chiamano coronati, ne' ricci capelluti, che ciuffoli da' nostri contadini son detti ; nei ricci legnosi del cerro, ne' ricci stellati della quercia, nelle galluzze della foglia del leccio si vede evidentissimamente, che la prima e principale intenzione della natura è formare dentro di quelle un animale volante ; vedendosi nel centro della gallozzola un uovo, che col crescere e col maturarsi di essa gallozzola va crescendo e maturando anch' egli, e cresce altresi a suo tempo quel verme, che nell' uovo si racchiude ; il qual verme, quando la gallozzola è finita di matu-rare e che è venuto il termine destinato al suo nascimento, diventa, di verme che era, una mosca. . . . Io vi confesso in-genuamente, che prima d'aver fatte queste mie esperienze intorno alla generazione degl' insetti mi dava a credere, o per dir meglio sospettava, che forse la gallozzola nascesse, perchè arrivando la mosca nel tempo della primavera, e facendo una piccolissima fessura ne' rami più teneri della quercia, in quella fessura nas-condesse uno de suoi semi, il quale fosse cagione che sbocciasse fuora la gallozzola ; e che mai non si vedessero galle o gallozzole o ricci o cornetti o calici o coccole, se non in que' rami, ne' quali le mosche avessero depositate le loro semenze ; e mi dava ad intendere, che le gallozzole fossero una malattia cagionata nelle

It is of great importance to apprehend Redi's position rightly; for the lines of thought he laid down for us are those upon which naturalists have been working ever since. Clearly, he held *Biogenesis* as against *Abiogenesis;* and I shall immediately proceed, in the first place, to inquire how far subsequent investigation has borne him out in so doing.

But Redi also thought that there were two modes of Biogenesis. By the one method, which is that of common and ordinary occurrence, the living parent gives rise to offspring which passes through the same cycle of changes as itself—like gives rise to like; and this has been termed *Homogenesis.* By the other mode, the living parent was supposed to give rise to offspring which passed through a totally different series of states from those exhibited by the parent, and did not return into the cycle of the parent; this is what ought to be called *Heterogenesis*, the offspring being altogether, and permanently, unlike the parent. The term Heterogenesis, however, has unfortunately been used in a different sense, and M. Milne-Edwards has therefore substituted for it *Xenogenesis*, which means the generation of something foreign. After discussing Redi's hypothesis of universal Biogenesis, then, I shall go

querce dalle punture delle mosche, in quella giusa stessa che dalle punture d'altri animaletti simiglievoli veggiamo crescere de' tumori ne' corpi degli animali."

on to ask how far the growth of science justifies his other hypothesis of Xenogenesis.

The progress of the hypothesis of Biogenesis was triumphant and unchecked for nearly a century. The application of the microscope to anatomy in the hands of Grew, Leeuwenhoek, Swammerdam, Lyonnet, Vallisnieri, Réaumur, and other illustrious investigators of nature of that day, displayed such a complexity of organisation in the lowest and minutest forms, and everywhere revealed such a prodigality of provision for their multiplication by germs of one sort or another, that the hypothesis of Abiogenesis began to appear not only untrue, but absurd; and, in the middle of the eighteenth century, when Needham and Buffon took up the question, it was almost universally discredited.[1]

But the skill of the microscope makers of the eighteenth century soon reached its limit. A microscope magnifying 400 diameters was a *chef d'œuvre* of the opticians of that day; and, at the same time, by no means trustworthy. But a magnifying power of 400 diameters, even when

---

[1] Needham, writing in 1750, says :—

"Les naturalistes modernes s'accordent unaninement à établir, comme une vérité certaine, que toute plante vient de sa sémence spécifique, tout animal d'un œuf ou de quelque chose d'analogue préexistant dans la plante, ou dans l'animal de même espèce qui l'a produit."—*Nouvelles Observations*, p. 169.

"Les naturalistes ont généralement cru que les animaux microscopiques étaient engendrés par des œufs transportés dans l'air, ou déposés dans des eaux dormantes par des insectes volans."—*Ibid.* p. 176.

definition reaches the exquisite perfection of our modern achromatic lenses, hardly suffices for the mere discernment of the smallest forms of life. A speck, only $\frac{1}{25}$th of an inch in diameter, has, at ten inches from the eye, the same apparent size as an object $\frac{1}{10000}$th of an inch in diameter, when magnified 400 times; but forms of living matter abound, the diameter of which is not more than $\frac{1}{40000}$th of an inch. A filtered infusion of hay, allowed to stand for two days, will swarm with living things among which, any which reaches the diameter of a human red blood-corpuscle, or about $\frac{1}{3200}$th of an inch, is a giant. It is only by bearing these facts in mind, that we can deal fairly with the remarkable statements and speculations put forward by Buffon and Needham in the middle of the eighteenth century.

When a portion of any animal or vegetable body is infused in water, it gradually softens and disintegrates; and, as it does so, the water is found to swarm with minute active creatures, the so-called Infusorial Animalcules, none of which can be seen, except by the aid of the microscope; while a large proportion belong to the category of smallest things of which I have spoken, and which must have looked like mere dots and lines under the ordinary microscopes of the eighteenth century.

Led by various theoretical considerations which

I cannot now discuss, but which looked promising enough in the lights of their time, Buffon and Needham doubted the applicability of Redi's hypothesis to the infusorial animalcules, and Needham very properly endeavoured to put the question to an experimental test. He said to himself, If these infusorial animalcules come from germs, their germs must exist either in the substance infused, or in the water with which the infusion is made, or in the superjacent air. Now the vitality of all germs is destroyed by heat. Therefore, if I boil the infusion, cork it up carefully, cementing the cork over with mastic, and then heat the whole vessel by heaping hot ashes over it, I must needs kill whatever germs are present. Consequently, if Redi's hypothesis hold good, when the infusion is taken away and allowed to cool, no animalcules ought to be developed in it; whereas, if the animalcules are not dependent on pre-existing germs, but are generated from the infused substance, they ought, by and by, to make their appearance. Needham found that, under the circumstances in which he made his experiments, animalcules always did arise in the infusions, when a sufficient time had elapsed to allow for their development.

In much of his work Needham was associated with Buffon, and the results of their experiments fitted in admirably with the great French naturalist's hypothesis of "organic molecules," according

to which, life is the indefeasible property of certain
indestructible molecules of matter, which exist in
all living things, and have inherent activities
by which they are distinguished from not living
matter. Each individual living organism is
formed by their temporary combination. They
stand to it in the relation of the particles of
water to a cascade, or a whirlpool; or to a mould,
into which the water is poured. The form of the
organism is thus determined by the reaction
between external conditions and the inherent
activities of the organic molecules of which it is
composed; and, as the stoppage of a whirlpool
destroys nothing but a form, and leaves the
molecules of the water, with all their inherent
activities intact, so what we call the death and
putrefaction of an animal, or of a plant, is merely
the breaking up of the form, or manner of association, of its constituent organic molecules, which
are then set free as infusorial animalcules.

It will be perceived that this doctrine is by no
means identical with *Abiogenesis*, with which it is
often confounded. On this hypothesis, a piece of
beef, or a handful of hay, is dead only in a limited
sense. The beef is dead ox, and the hay is dead
grass; but the " organic molecules " of the beef or
the hay are not dead, but are ready to manifest
their vitality as soon as the bovine or herbaceous
shrouds in which they are imprisoned are rent by
the macerating action of water. The hypothesis

R 2

therefore must be classified under Xenogenesis, rather than under Abiogenesis. Such as it was, I think it will appear, to those who will be just enough to remember that it was propounded before the birth of modern chemistry, and of the modern optical arts, to be a most ingenious and suggestive speculation.

But the great tragedy of Science—the slaying of a beautiful hypothesis by an ugly fact—which is so constantly being enacted under the eyes of philosophers, was played, almost immediately, for the benefit of Buffon and Needham.

Once more, an Italian, the Abbé Spallanzani, a worthy successor and representative of Redi in his acuteness, his ingenuity, and his learning, subjected the experiments and the conclusions of Needham to a searching criticism. It might be true that Needham's experiments yielded results such as he had described, but did they bear out his arguments? Was it not possible, in the first place, he had not completely excluded the air by his corks and mastic? And was it not possible, in the second place, that he had not sufficiently heated his infusions and the superjacent air? Spallanzani joined issue with the English naturalist on both these pleas, and he showed that if, in the first place, the glass vessels in which the infusions were contained were hermetically sealed by fusing their necks, and if, in the second place, they were exposed to the temperature of boiling water for

three-quarters of an hour,[1] no animalcules ever
made their appearance within them. It must be
admitted that the experiments and arguments of
Spallanzani furnish a complete and a crushing
reply to those of Needham. But we all too often
forget that it is one thing to refute a proposition,
and another to prove the truth of a doctrine which,
implicitly or explicitly, contradicts that proposition ; and the advance of science soon showed
that though Needham might be quite wrong, it
did not follow that Spallanzani was quite right.

Modern Chemistry, the birth of the latter half
of the eighteenth century, grew apace, and soon
found herself face to face with the great problems
which biology had vainly tried to attack without
her help. The discovery of oxygen led to the laying of the foundations of a scientific theory of
respiration, and to an examination of the marvellous interactions of organic substances with
oxygen. The presence of free oxygen appeared to
be one of the conditions of the existence of life,
and of those singular changes in organic matters
which are known as fermentation and putrefaction.
The question of the generation of the infusory
animalcules thus passed into a new phase. For
what might not have happened to the organic
matter of the infusions, or to the oxygen of the
air, in Spallanzani's experiments ? What security
was there that the development of life which ought

[1] See Spallanzani, *Opere*, vi. pp. 42 and 51.

to have taken place had not been checked or pre-vented by these changes?

The battle had to be fought again. It was need-ful to repeat the experiments under conditions which would make sure that neither the oxygen of the air, nor the composition of the organic matter, was altered in such a manner as to interfere with the existence of life.

Schulze and Schwann took up the question from this point of view in 1836 and 1837. The passage of air through red-hot glass tubes, or through strong sulphuric acid, does not alter the propor-tion of its oxygen, while it must needs arrest, or destroy, any organic matter which may be con-tained in the air. These experimenters, therefore, contrived arrangements by which the only air which should come into contact with a boiled in-fusion should be such as had either passed through red-hot tubes or through strong sulphuric acid. The result which they obtained was that an in-fusion so treated developed no living things, while, if the same infusion was afterwards exposed to the air, such things appeared rapidly and abundantly. The accuracy of these experiments has been alternately denied and affirmed. Supposing them to be accepted, however, all that they really proved was that the treatment to which the air was subjected destroyed *something* that was essential to the development of life in the infusion. This "something" might be gaseous, fluid, or solid;

that it consisted of germs remained only an hypothesis of greater or less probability.

Contemporaneously with these investigations a
remarkable discovery was made by Cagniard de la
Tour. He found that common yeast is composed of a vast accumulation of minute plants.
The fermentation of must, or of wort, in the
fabrication of wine and of beer, is always accompanied by the rapid growth and multiplication of
these *Torulæ*. Thus, fermentation, in so far as it
was accompanied by the development of microscopical organisms in enormous numbers, became
assimilated to the decomposition of an infusion of
ordinary animal or vegetable matter ; and it was
an obvious suggestion that the organisms were, in
some way or other, the causes both of fermentation
and of putrefaction. The chemists, with Berzelius
and Liebig at their head, at first laughed this idea
to scorn ; but in 1843, a man then very young,
who has since performed the unexampled feat of
attaining to high eminence alike in Mathematics,
Physics, and Physiology—I speak of the illustrious
Helmholtz— reduced the matter to the test of
experiment by a method alike elegant and conclusive. Helmholtz separated a putrefying or a
fermenting liquid from one which was simply
putrescible or fermentable by a membrane which
allowed the fluids to pass through and become
intermixed, but stopped the passage of solids.
The result was, that while the putrescible or the

fermentable liquids became impregnated with the results of the putrescence or fermentation which was going on on the other side of the membrane, they neither putrefied (in the ordinary way) nor fermented; nor were any of the organisms which abounded in the fermenting or putrefying liquid generated in them. Therefore the cause of the development of these organisms must lie in something which cannot pass through membranes; and as Helmholtz's investigations were long antecedent to Graham's researches upon colloids, his natural conclusion was that the agent thus intercepted must be a solid material. In point of fact, Helmholtz's experiments narrowed the issue to this: that which excites fermentation and putrefaction, and at the same time gives rise to living forms in a fermentable or putrescible fluid, is not a gas and is not a diffusible fluid; therefore it is either a colloid, or it is matter divided into very minute solid particles.

The researches of Schroeder and Dusch in 1854, and of Schroeder alone, in 1859, cleared up this point by experiments which are simply refinements upon those of Redi. A lump of cotton-wool is, physically speaking, a pile of many thicknesses of a very fine gauze, the fineness of the meshes of which depends upon the closeness of the compression of the wool. Now, Schroeder and Dusch found, that, in the case of all the putrefiable materials which they used (except milk and yolk

of egg), an infusion boiled, and then allowed to
come into contact with no air but such as had
been filtered through cotton-wool, neither putrefied, nor fermented, nor developed living forms.
It is hard to imagine what the fine sieve formed
by the cotton-wool could have stopped except
minute solid particles. Still the evidence was
incomplete until it had been positively shown,
first, that ordinary air does contain such particles ;
and, secondly, that filtration through cotton-wool
arrests these particles and allows only physically
pure air to pass. This demonstration has been
furnished within the last year by the remarkable
experiments of Professor Tyndall. It has been a
common objection of Abiogenists that, if the
doctrine of Biogeny is true, the air must be thick
with germs ; and they regard this as the height of
absurdity. But nature occasionally is exceedingly
unreasonable, and Professor Tyndall has proved
that this particular absurdity may nevertheless be
a reality. He has demonstrated that ordinary air
is no better than a sort of stirabout of excessively
minute solid particles ; that these particles are
almost wholly destructible by heat ; and that they
are strained off, and the air rendered optically
pure, by its being passed through cotton-wool.

It remains yet in the order of logic, though
not of history, to show that among these solid
destructible particles, there really do exist germs
capable of giving rise to the development of living

forms in suitable menstrua. This piece of work was done by M. Pasteur in those beautiful researches which will ever render his name famous; and which, in spite of all attacks upon them, appear to me now, as they did seven years ago,[1] to be models of accurate experimentation and logical reasoning. He strained air through cotton-wool, and found, as Schroeder and Dusch had done, that it contained nothing competent to give rise to the development of life in fluids highly fitted for that purpose. But the important further links in the chain of evidence added by Pasteur are three. In the first place he subjected to microscopic examination the cotton-wool which had served as strainer, and found that sundry bodies clearly recognisable as germs, were among the solid particles strained off. Secondly, he proved that these germs were competent to give rise to living forms by simply sowing them in a solution fitted for their development. And, thirdly, he showed that the incapacity of air strained through cotton-wool to give rise to life, was not due to any occult change effected in the constituents of the air by the wool, by proving that the cotton-wool might be dispensed with altogether, and perfectly free access left between the exterior air and that in the experimental flask. If the neck of the flask is drawn out into a tube and bent downwards; and

[1] *Lectures to Working Men on the Causes of the Phenomena of Organic Nature*, 1863. (See Vol. II. of these Essays.)

if, after the contained fluid has been carefully
boiled, the tube is heated sufficiently to destroy
any germs which may be present in the air which
enters as the fluid cools, the apparatus may be
left to itself for any time and no life will appear
in the fluid. The reason is plain. Although there
is free communication between the atmosphere
laden with germs and the germless air in the flask,
contact between the two takes place only in the
tube; and as the germs cannot fall upwards, and
there are no currents, they never reach the interior
of the flask. But if the tube be broken short off
where it proceeds from the flask, and free access
be thus given to germs falling vertically out of
the air, the fluid, which has remained clear and
desert for months, becomes, in a few days, turbid
and full of life.

These experiments have been repeated over and
over again by independent observers with entire
success; and there is one very simple mode of
seeing the facts for one's self, which I may as well
describe.

Prepare a solution (much used by M. Pasteur,
and often called " Pasteur's solution ") composed
of water with tartrate of ammonia, sugar, and
yeast-ash dissolved therein.[1] Divide it into three
portions in as many flasks; boil all three for a

---

[1] Infusion of hay treated in the same way yields similar
results ; but as it contains organic matter, the argument which
follows cannot be based upon it.

quarter of an hour; and, while the steam is passing
out, stop the neck of one with a large plug of
cotton-wool, so that this also may be thoroughly
steamed. Now set the flasks aside to cool, and,
when their contents are cold, add to one of the open
ones a drop of filtered infusion of hay which has
stood for twenty-four hours, and is consequently
full of the active and excessively minute organisms
known as *Bacteria*. In a couple of days of ordinary
warm weather the contents of this flask will be
milky from the enormous multiplication of *Bacteria*.
The other flask, open and exposed to the air, will,
sooner or later, become milky with *Bacteria*, and
patches of mould may appear in it; while the liquid
in the flask, the neck of which is plugged with
cotton-wool, will remain clear for an indefinite
time. I have sought in vain for any explanation
of these facts, except the obvious one, that the air
contains germs competent to give rise to *Bacteria*,
such as those with which the first solution has
been knowingly and purposely inoculated, and to
the mould-*Fungi*. And I have not yet been able
to meet with any advocate of Abiogenesis who
seriously maintains that the atoms of sugar, tartrate of ammonia, yeast-ash, and water, under no influence but that of free access of air and the ordinary
temperature, re-arrange themselves and give rise
to the protoplasm of *Bacterium*. But the alternative is to admit that these *Bacteria* arise from
germs in the air; and if they are thus propagated,

the burden of proof that other like forms are generated in a different manner, must rest with the assertor of that proposition.

To sum up the effect of this long chain of evidence :—

It is demonstrable that a fluid eminently fit for the development of the lowest forms of life, but which contains neither germs, nor any protein compound, gives rise to living things in great abundance if it is exposed to ordinary air; while no such development takes place, if the air with which it is in contact is mechanically freed from the solid particles which ordinarily float in it, and which may be made visible by appropriate means

It is demonstrable that the great majority of these particles are destructible by heat, and that some of them are germs, or living particles, capable of giving rise to the same forms of life as those which appear when the fluid is exposed to un-purified air.

It is demonstrable that inoculation of the experimental fluid with a drop of liquid known to contain living particles gives rise to the same phenomena as exposure to unpurified air.

And it is further certain that these living particles are so minute that the assumption of their suspension in ordinary air presents not the slightest difficulty. On the contrary, considering their lightness and the wide diffusion of the organisms which produce them, it is impossible to conceive that they should not be suspended in the atmosphere in myriads.

Thus the evidence, direct and indirect, in favour of *Biogenesis* for all known forms of life must, I think, be admitted to be of great weight.

On the other side, the sole assertions worthy of attention are that hermetically sealed fluids, which have been exposed to great and long-continued heat, have sometimes exhibited living forms of low organisation when they have been opened.

The first reply that suggests itself is the probability that there must be some error about these experiments, because they are performed on an enormous scale every day with quite contrary results. Meat, fruits, vegetables, the very materials of the most fermentable and putrescible infusions, are preserved to the extent, I suppose I may say, of thousands of tons every year, by a method which is a mere application of Spallanzani's experiment. The matters to be preserved are well boiled in a tin case provided with a small hole, and this hole is soldered up when all the air in the case has been replaced by steam. By this method they may be kept for years without putrefying, fermenting, or getting mouldy. Now this is not because oxygen is excluded, inasmuch as it is now proved that free oxygen is not necessary for either fermentation or putrefaction. It is not because the tins are exhausted of air, for

Vibriones and *Bacteria* live, as Pasteur has shown, without air or free oxygen. It is not because the boiled meats or vegetables are not putrescible or fermentable, as those who have had the misfortune to be in a ship supplied with unskilfully closed tins well know. What is it, therefore, but the exclusion of germs ? I think that Abiogenists are bound to answer this question before they ask us to consider new experiments of precisely the same order.

And in the next place, if the results of the experiments I refer to are really trustworthy, it by no means follows that Abiogenesis has taken place. The resistance of living matter to heat is known to vary within considerable limits, and to depend, to some extent, upon the chemical and physical qualities of the surrounding medium. But if, in the present state of science, the alternative is offered us,—either germs can stand a greater heat than has been supposed, or the molecules of dead matter, for no valid or intelligible reason that is assigned, are able to re-arrange themselves into living bodies, exactly such as can be demonstrated to be frequently produced in another way,—I cannot understand how choice can be, even for a moment, doubtful.

But though I cannot express this conviction of mine too strongly, I must carefully guard myself against the supposition that I intend to suggest that no such thing as Abiogenesis ever has taken place in the past, or ever will take place in the future.   With organic chemistry, molecular physics, and physiology yet in their infancy, and every day making prodigious strides, I think it would be the height of presumption for any man to say that the conditions under which matter assumes the properties we call "vital" may not, some day, be artificially brought together.   All I feel justified in affirming is, that I see no reason for believing that the feat has been performed yet.

And looking back through the prodigious vista of the past, I find no record of the commencement of life, and therefore I am devoid of any means of forming a definite conclusion as to the conditions of its appearance.   Belief, in the scientific sense of the word, is a serious matter, and needs strong foundations.   To say, therefore, in the admitted absence of evidence, that I have any belief as to the mode in which the existing forms of life have originated, would be using words in a wrong sense. But expectation is permissible where belief is not; and if it were given me to look beyond the abyss of geologically recorded time to the still more remote period when the earth was passing through physical and chemical conditions, which it can no more see again than a man can recall his infancy, I should expect to be a witness of the evolution of living protoplasm from not living matter.   I should expect to see it appear under forms of great simplicity, endowed, like existing fungi, with the power of determining the formation of new protoplasm from such matters as ammonium carbonates, oxalates and tartrates, alkaline and earthy phosphates, and water, without the aid of light. That is the expectation to which analogical reasoning leads me ; but I beg you once more to recollect that I have no right to call my opinion anything but an act of philosophical faith.

So much for the history of the progress of Redi's great doctrine of Biogenesis, which appears to me, with the limitations I have expressed, to be victorious along the whole line at the present day.

As regards the second problem offered to us by Redi, whether Xenogenesis obtains, side by side with Homogenesis,—whether, that is, there exist not only the ordinary living things, giving rise to offspring which run through the same cycle as themselves, but also others, producing offspring which are of a totally different character from themselves,—the researches of two centuries have led to a different result. That the grubs found in galls are no product of the plants on which the galls grow, but are the result of the introduction of the eggs of insects into the substance of these plants, was made out by Vallisnieri, Réaumur, and others, before the end of the first half of the eighteenth century. The tapeworms, bladderworms, and flukes continued to be a stronghold of the advocates of Xenogenesis for a much longer period. Indeed, it is only within the last thirty years that the splendid patience of Von Siebold, Van Beneden, Leuckart, Küchenmeister, and other helminthologists, has succeeded in tracing every such parasite, often through the strangest wanderings and metamorphoses, to an egg derived from a parent, actually or potentially like itself; and the tendency of inquiries elsewhere has all been in the same direction. A plant may throw off bulbs, but these, sooner or later, give rise to seeds or spores, which develop into the original form. A polype may give rise to Medusæ, or a pluteus to an Echinoderm, but the Medusa and the Echinoderm give rise to eggs which produce polypes or plutei, and they are therefore only stages in the cycle of life of the species.

But if we turn to pathology, it offers us some remarkable approximations to true Xenogenesis.

As I have already mentioned, it has been known since the time of Vallisnieri and of Réaumur, that galls in plants, and tumours in cattle, are caused by insects, which lay their eggs in those parts of the animal or vegetable frame of which these morbid structures are outgrowths. Again, it is a matter of familiar experience to everybody that mere pressure on the skin will give rise to a corn. Now the gall, the tumour, and the corn are parts of the living body, which have become, to a certain degree, independent and distinct organisms. Under the influence of certain external conditions, elements of the body, which should have developed in due subordination to its general plan, set up for themselves and apply the nourishment which they receive to their own purposes.

From such innocent productions as corns and warts, there are all gradations to the serious tumours which, by their mere size and the mechanical obstruction they cause, destroy the organism out of which they are developed ; while, finally, in those terrible structures known as cancers, the abnormal growth has acquired powers of reproduction and multiplication, and is only morphologically distinguishable from the parasitic worm, the life of which is neither more nor less closely bound up with that of the infested organism.

If there were a kind of diseased structure, the histological elements of which were capable of maintaining a separate and independent existence out of the body, it seems to me that the shadowy boundary between morbid growth and Xeno-genesis would be effaced. And I am inclined to think that the progress of discovery has almost brought us to this point already. I have been favoured by Mr. Simon with an early copy of the last published of the valuable " Reports on the

Public Health," which, in his capacity of their
medical officer, he annually presents to the Lords
of the Privy Council. The appendix to this
report contains an introductory essay " On the
Intimate Pathology of Contagion," by Dr. Burdon-
Sanderson, which is one of the clearest, most
comprehensive, and well-reasoned discussions of a
great question which has come under my notice
for a long time. I refer you to it for details and
for the authorities for the statements I am about
to make.

You are familiar with what happens in vaccina-
tion. A minute cut is made in the skin, and an
infinitesimal quantity of vaccine matter is inserted
into the wound. Within a certain time a vesicle
appears in the place of the wound, and the fluid
which distends this vesicle is vaccine matter, in
quantity a hundred or a thousandfold that which
was originally inserted. Now what has taken
place in the course of this operation ? Has the
vaccine matter, by its irritative property, produced
a mere blister, the fluid of which has the same
irritative property ? Or does the vaccine matter
contain living particles, which have grown and
multiplied where they have been planted ? The
observations of M. Chauveau, extended and con-
firmed by Dr. Sanderson himself, appear to leave
no doubt upon this head. Experiments, similar
in principle to those of Helmholtz on fermentation
and putrefaction, have proved that the active

element in the vaccine lymph is non-diffusible, and consists of minute particles not exceeding $\frac{1}{20000}$th of an inch in diameter, which are made visible in the lymph by the microscope. Similar experiments have proved that two of the most destructive of epizootic diseases, sheep-pox and glanders, are also dependent for their existence and their propagation upon extremely small living solid particles, to which the title of *microzymes* is applied. An animal suffering under either of these terrible diseases is a source of infection and contagion to others, for precisely the same reason as a tub of fermenting beer is capable of propagating its fermentation by "infection," or "contagion," to fresh wort. In both cases it is the solid living particles which are efficient; the liquid in which they float, and at the expense of which they live, being altogether passive.

Now arises the question, are these microzymes the results of *Homogenesis*, or of *Xenogenesis*? are they capable, like the *Torulæ* of yeast, of arising only by the development of pre-existing germs? or may they be, like the constituents of a nut-gall, the results of a modification and individualisation of the tissues of the body in which they are found, resulting from the operation of certain conditions? Are they parasites in the zoological sense, or are they merely what Virchow has called "heterologous growths"? It is obvious that this question has the most profound importance, whether we look at it from a practical or from a theoretical point of view. A parasite may be stamped out by destroying its germs, but a pathological product can only be annihilated by removing the conditions which give rise to it.

It appears to me that this great problem will have to be solved for each zymotic disease separately, for analogy cuts two ways. I have dwelt upon the analogy of pathological modification, which is in favour of the xenogenetic origin of microzymes; but I must now speak of the equally strong analogies in favour of the origin of such pestiferous particles by the ordinary process of the generation of like from like.

It is, at present, a well-established fact that certain diseases, both of plants and of animals, which have all the characters of contagious and infectious epidemics, are caused by minute organisms. The smut of wheat is a well-known instance of such a disease, and it cannot be doubted that the grape-disease and the potato-disease fall under the same category. Among animals, insects are wonderfully liable to the ravages of contagious and infectious diseases caused by microscopic *Fungi*.

In autumn, it is not uncommon to see flies motionless upon a window-pane, with a sort of magic circle, in white, drawn round them. On microscopic examination, the magic circle is found to consist of innumerable spores, which have been thrown off in all directions by a minute fungus called *Empusa muscæ*, the spore-forming filaments of which stand out like a pile of velvet from the body of the fly. These spore-forming filaments are connected with others which fill the interior of the fly's body like so much fine wool, having eaten away and destroyed the creature's viscera. This is the full-grown condition of the *Empusa*. If traced back to its earliest stages, in flies which are still active, and to all appearance healthy, it is found to exist in the form of minute corpuscles which float in the blood of the fly. These multiply and lengthen into filaments, at the expense of the fly's substance; and when they have at last killed the patient, they grow out of its body and give off spores. Healthy flies shut up with diseased ones catch this mortal disease, and perish like the others. A most competent observer, M. Cohn, who studied the development of the *Empusa* very carefully, was utterly unable to discover in what manner the smallest germs of the *Empusa* got into the fly. The spores could not be made to give rise to such germs by cultivation; nor were such germs discoverable in the air, or in the food of the fly. It looked exceedingly like a case of Abiogenesis, or, at any rate, of Xenogenesis; and it is only quite recently that the real course of events has been made out. It has been ascertained, that when one of the spores falls upon the body of a fly, it begins to germinate, and sends out a process which bores its way through the fly's skin; this, having reached the interior cavities of its body, gives off the minute floating corpuscles which are the earliest stage of the *Empusa*. The disease is "contagious," because a healthy fly coming in contact with a diseased one, from which the spore-bearing filaments protrude, is pretty sure to carry off a spore or two. It is "infectious" because the spores become scattered about all sorts of matter in the neighbourhood of the slain flies.

The silkworm has long been known to be subject to a very fatal and infectious disease called the *Muscardine*. Audouin transmitted it by inoculation. This disease is entirely due to the development of a fungus, *Botrytis Bassiana*, in the body of the caterpillar; and its contagiousness and infectiousness are accounted for in the same way as those of the fly-disease. But, of late years, a still more serious epizootic has appeared among the silkworms; and I may mention a few facts which will give you some conception of the gravity of the injury which it has inflicted on France alone.

The production of silk has been for centuries an important branch of industry in Southern France, and in the year 1853 it had attained such a magnitude that the annual produce of the French sericulture was estimated to amount to a tenth of that of the whole world, and represented a money-value of 117,000,000 francs, or nearly five millions sterling. What may be the sum which would represent the money-value of all the industries connected with the working up of the raw silk thus produced, is more than I can pretend to estimate. Suffice it to say, that the city of Lyons is built upon French silk as much as Manchester was upon American cotton before the civil war.

Silkworms are liable to many diseases; and, even before 1853, a peculiar epizootic, frequently accompanied by the appearance of dark spots upon the skin (whence the name of " Pébrine " which it has received), had been noted for its mortality. But in the years following 1853 this malady broke out with such extreme violence, that, in 1858, the silk-crop was reduced to a third of the amount which it had reached in 1853; and, up till within the last year or two, it has never attained half the yield of 1853. This means not only that the great number of people engaged in silk growing are some thirty millions sterling poorer than they might have been; it means not only that high prices have had to be paid for imported silkworm eggs, and that, after investing his money in them, in paying for mulberry-leaves and for attendance, the cultivator has constantly seen his silkworms perish and himself plunged in ruin; but it means that the looms of

Lyons have lacked employment, and that, for years, enforced idleness and misery have been the portion of a vast population which, in former days, was industrious and well-to-do.

In 1858 the gravity of the situation caused the French Academy of Sciences to appoint Commissioners, of whom a distinguished naturalist, M. de Quatrefages, was one, to inquire into the nature of this disease, and, if possible, to devise some means of staying the plague. In reading the Report [1] made by M. de Quatrefages in 1859, it is exceedingly interesting to observe that his elaborate study of the Pébrine forced the conviction upon his mind that, in its mode of occurrence and propagation, the disease of the silkworm is, in every respect, comparable to the cholera among mankind. But it differs from the cholera, and so far is a more formidable malady, in being hereditary, and in being, under some circumstances, contagious as well as infectious.

The Italian naturalist, Filippi, discovered in the blood of the silkworms affected by this strange disorder a multitude of cylindrical corpuscles, each about $\frac{1}{6000}$th of an inch long. These have been carefully studied by Lebert, and named by him *Panhistophyton;* for the reason that in subjects in which the disease is strongly developed, the corpuscles swarm in every tissue and organ of the body, and even pass into the undeveloped eggs of

[1] *Études sur les Maladies actuelles des Vers à Soie,* p. 53.

the female moth. But are these corpuscles causes, or mere concomitants, of the disease ? Some naturalists took one view and some another ; and it was not until the French Government, alarmed by the continued ravages of the malady, and the inefficiency of the remedies which had been suggested, despatched M. Pasteur to study it, that the question received its final settlement ; at a great sacrifice, not only of the time and peace of mind of that eminent philosopher, but, I regret to have to add, of his health.

But the sacrifice has not been in vain. It is now certain that this devastating, cholera-like, Pébrine, is the effect of the growth and multiplication of the *Panhistophyton* in the silkworm. It is contagious and infectious, because the corpuscles of the *Panhistophyton* pass away from the bodies of the diseased caterpillars, directly or indirectly, to the alimentary canal of healthy silkworms in their neighbourhood ; it is hereditary because the corpuscles enter into the eggs while they are being formed, and consequently are carried within them when they are laid ; and for this reason, also, it presents the very singular peculiarity of being inherited only on the mother's side. There is not a single one of all the apparently capricious and unaccountable phenomena presented by the Pébrine, but has received its explanation from the fact that the disease is the result of the presence of the microscopic organism, *Panhistophyton*.

Such being the facts with respect to the Pébrine, what are the indications as to the method of preventing it ? It is obvious that this depends upon the way in which the *Panhistophyton* is generated. If it may be generated by Abiogenesis, or by Xenogenesis, within the silkworm or its moth, the extirpation of the disease must depend upon the prevention of the occurrence of the conditions under which this generation takes place. But if, on the other hand, the *Panhistophyton* is an independent organism, which is no more generated by the silkworm than the mistletoe is generated by the apple-tree or the oak on which it grows, though it may need the silkworm for its development in the same way as the mistletoe needs the tree, then the indications are totally different. The sole thing to be done is to get rid of and keep away the germs of the *Panhistophyton*. As might be imagined, from the course of his previous investigations, M. Pasteur was led to believe that the latter was the right theory ; and, guided by that theory, he has devised a method of extirpating the disease, which has proved to be completely successful wherever it has been properly carried out.

There can be no reason, then, for doubting that, among insects, contagious and infectious diseases, of great malignity, are caused by minute organisms which are produced from pre-existing germs, or by homogenesis ; and there is no reason, that I know of, for believing that what happens in insects may not take place in the highest animals. Indeed, there is already strong evidence that some diseases of an extremely malignant and fatal character to which man is subject, are as much the work of minute organisms as is the Pébrine. I refer for this evidence to the very striking facts adduced by Professor Lister in his various well-known publications on the antiseptic method of treatment. It appears to me impossible to rise from the perusal of those publications without a strong conviction that the lamentable mortality which so frequently dogs the footsteps of the most skilful operator, and those deadly consequences of wounds and injuries which seem to haunt the very walls of great hospitals, and are, even now, destroying more men than die of bullet or bayonet, are due to the importation of minute organisms into wounds, and their increase and multiplication; and that the surgeon who saves most lives will be he who best works out the practical consequences of the hypothesis of Redi.

I commenced this Address by asking you to follow me in an attempt to trace the path which has been followed by a scientific idea, in its long and slow progress from the position of a probable hypothesis to that of an established law of nature. Our survey has not taken us into very attractive regions; it has lain, chiefly, in a land flowing with the abominable, and peopled with mere grubs and mouldiness. And it may be imagined with what smiles and shrugs, practical and serious contemporaries of Redi and of Spallanzani may have commented on the waste of their high abilities in toiling at the solution of problems which, though curious enough in themselves, could be of no conceivable utility to mankind.

Nevertheless, you will have observed that before we had travelled very far upon our road, there appeared, on the right hand and on the left, fields laden with a harvest of golden grain, immediately convertible into those things which the most solidly practical men will admit to have value—viz., money and life.

The direct loss to France caused by the Pébrine in seventeen years cannot be estimated at less than fifty millions sterling; and if we add to this what Redi's idea, in Pasteur's hands, has done for the wine-grower and for the vinegar-maker, and try to capitalise its value, we shall find that it will go a long way towards repairing the money losses caused by the frightful and calamitous war of this autumn. And as to the equivalent of Redi's thought in life, how can we over-estimate the value of that knowledge of the nature of epidemic and epizootic diseases, and consequently of the means of checking, or eradicating them, the dawn of which has assuredly commenced ?

Looking back no further than ten years, it is possible to select three (1863, 1864, and 1869) in which the total number of deaths from scarlet-fever alone amounted to ninety thousand. That is the return of killed, the maimed and disabled being left out of sight. Why, it is to be hoped that the list of killed in the present bloodiest of all wars will not amount to more than this! But the facts which I have placed before you must leave the least sanguine without a doubt that the nature and the causes of this scourge will, one day, be as well understood as those of the Pébrine are now; and that the long-suffered massacre of our innocents will come to an end.

And thus mankind will have one more admonition that "the people perish for lack of knowledge"; and that the alleviation of the miseries, and the promotion of the welfare, of men must be sought, by those who will not lose their pains, in that diligent, patient, loving study of all the multitudinous aspects of Nature, the results of which constitute exact knowledge, or Science. It is the justification and the glory of this great meeting that it is gathered together for no other object than the advancement of the moiety of science which deals with those phenomena of nature which we call physical. May its endeavours be crowned with a full measure of success!

IX

GEOLOGICAL CONTEMPORANEITY AND PERSISTENT TYPES OF LIFE

[1862]

MERCHANTS occasionally go through a wholesome, though troublesome and not always satisfactory, process which they term " taking stock." After all the excitement of speculation, the pleasure of gain, and the pain of loss, the trader makes up his mind to face facts and to learn the exact quantity and quality of his solid and reliable possessions.

The man of science does well sometimes to imitate this procedure : and, forgetting for the time the importance of his own small winnings, to re-examine the common stock in trade, so that he may make sure how far the stock of bullion in the cellar—on the faith of whose existence so much paper has been circulating—is really the solid gold of truth.

The Anniversary Meeting of the Geological

Society seems to be an occasion well suited for an undertaking of this kind—for an inquiry, in fact, into the nature and value of the present results of palæontological investigation; and the more so, as all those who have paid close attention to the late multitudinous discussions in which palæontology is implicated, must have felt the urgent necessity of some such scrutiny.

First in order, as the most definite and unquestionable of all the results of palæontology, must be mentioned the immense extension and impulse given to botany, zoology, and comparative anatomy, by the investigation of fossil remains. Indeed, the mass of biological facts has been so greatly increased, and the range of biological speculation has been so vastly widened, by the researches of the geologist and palæontologist, that it is to be feared there are naturalists in existence who look upon geology as Brindley regarded rivers. " Rivers," said the great engineer, " were made to feed canals ; " and geology, some seem to think, was solely created to advance comparative anatomy.

Were such a thought justifiable, it could hardly expect to be received with favour by this assembly. But it is not justifiable. Your favourite science has her own great aims independent of all others ; and if, notwithstanding her steady devotion to her own progress, she can scatter such rich alms among her sisters, it should be remembered that her charity is of the sort that does not impoverish, but " blesseth him that gives and him that takes."

Regard the matter as we will, however, the facts remain. Nearly 40,000 species of animals and plants have been added to the Systema Naturæ by palæontological research. This is a living population equivalent to that of a new continent in mere number ; equivalent to that of a new hemisphere, if we take into account the small population of insects as yet found fossil, and the large proportion and peculiar organisation of many of the Vertebrata.

But, beyond this, it is perhaps not too much to say that, except for the necessity of interpreting palæontological facts, the laws of distribution would have received less careful study; while few comparative anatomists (and those not of the first order) would have been induced by mere love of detail, as such, to study the minutiæ of osteology, were it not that in such minutiæ lie the only keys to the most interesting riddles offered by the extinct animal world.

These assuredly are great and solid gains. Surely it is matter for no small congratulation that in half a century (for palæontology, though it dawned earlier, came into full day only with Cuvier) a subordinate branch of biology should have doubled the value and the interest of the whole group of sciences to which it belongs.

But this is not all. Allied with geology, palæontology has established two laws of inestimable importance : the first, that one and the same area of the earth's surface has been successively occupied by very different kinds of living beings ; the second, that the order of succession established in one locality holds good, approximately, in all.

The first of these laws is universal and irreversible ; the second is an induction from a vast number of observations, though it may possibly, and even probably, have to admit of exceptions. As a consequence of the second law, it follows that a peculiar relation frequently subsists between series of strata containing organic remains, in different localities. The series resemble one another not only in virtue of a general resemblance of the organic remains in the two, but also in virtue of a resemblance in the order and character of the serial succession in each. There is a resemblance of arrangement ; so that the separate terms of each series, as well as the whole series, exhibit a correspondence.

Succession implies time ; the lower members of an undisturbed series of sedimentary rocks are certainly older than the upper ; and when the notion of age was once introduced as the equivalent of succession, it was no wonder that correspondence in succession came to be looked upon as a correspondence in age, or "contemporaneity." And, indeed, so long as relative age only is spoken of, correspondence in succession *is* correspondence in age ; it is *relative* contemporaneity.

But it would have been very much better for geology if so loose and ambiguous a word as "contemporaneous" had been excluded from her terminology, and if, in its stead, some term expressing similarity of serial relation, and excluding the notion of time altogether, had been employed to denote correspondence in position in two or more series of strata.

In anatomy, where such correspondence of position has constantly to be spoken of, it is denoted by the word "homology" and its derivatives ; and for Geology (which after all is only the anatomy and physiology of the earth) it might be well to invent some single word, such as "homotaxis" (similarity of order), in order to express an essentially similar idea. This, however, has not been done, and most probably the inquiry will at once be made—To what end burden science with a new and strange term in place of one old, familiar, and part of our common language ?

The reply to this question will become obvious as the inquiry into the results of palæontology is pushed further.

Those whose business it is to acquaint themselves specially with the works of palæontologists, in fact, will be fully aware that very few, if any, would rest satisfied with such a statement of the conclusions

of their branch of biology as that which has just
been given.

Our standard repertories of palæontology profess
to teach us far higher things—to disclose the
entire succession of living forms upon the surface
of the globe; to tell us of a wholly different distribution of climatic conditions in ancient times;
to reveal the character of the first of all living
existences; and to trace out the law of progress
from them to us.

It may not be unprofitable to bestow on these
professions a somewhat more critical examination
than they have hitherto received, in order to
ascertain how far they rest on an irrefragable
basis; or whether, after all, it might not be well
for palæontologists to learn a little more carefully
that scientific "ars artium," the art of saying "I
don't know."   And to this end let us define somewhat more exactly the extent of these pretensions
of palæontology.

Every one is aware that Professor Bronn's "Untersuchungen" and Professor Pictet's "Traité de
Paléontologie" are works of standard authority,
familiarly consulted by every working palæontologist.   It is desirable to speak of these excellent
books, and of their distinguished authors, with the
utmost respect, and in a tone as far as possible removed from carping criticism; indeed, if they are
specially cited in this place, it is merely in justification of the assertion that the following proposi-

tions, which may be found implicitly, or explicitly. in the works in question, are regarded by the mass of palæontologists and geologists, not only on the Continent but in this country, as expressing some of the best-established results of palæontology. Thus :—

Animals and plants began their existence together, not long after the commencement of the deposition of the sedimentary rocks; and then succeeded one another, in such a manner, that totally distinct faunæ and floræ occupied the whole surface of the earth, one after the other, and during distinct epochs of time.

A geological formation is the sum of all the strata deposited over the whole surface of the earth during one of these epochs: a geological fauna or flora is the sum of all the species of animals or plants which occupied the whole surface of the globe, during one of these epochs.

The population of the earth's surface was at first very similar in all parts, and only from the middle of the Tertiary epoch onwards, began to show a distinct distribution in zones.

The constitution of the original population, as well as the numerical proportions of its members, indicates a warmer and, on the whole, somewhat tropical climate, which remained tolerably equable throughout the year.   The subsequent distribution of living beings in zones is the result of a gradual lowering of the general temperature, which first began to be felt at the poles.

It is not now proposed to inquire whether these doctrines are true or false; but to direct your attention to a much simpler though very essential preliminary question—What is their logical basis ? what are the fundamental assumptions upon which they all logically depend ? and what is the evidence on which those fundamental propositions demand our assent ?

These assumptions are two : the first, that the commencement of the geological record is coëval with the commencement of life on the globe; the second, that geological contemporaneity is the same thing as chronological synchrony. Without the first of these assumptions there would of course be no ground for any statement respecting the commencement of life; without the second, all the other statements cited, every one of which implies a knowledge of the state of different parts of the earth at one and the same time, will be no less devoid of demonstration.

The first assumption obviously rests entirely on negative evidence. This is, of course, the only evidence that ever can be available to prove the commencement of any series of phenomena; but, at the same time, it must be recollected that the value of negative evidence depends entirely on the amount of positive corroboration it receives. If A.B.

wishes to prove an *alibi*, it is of no use for him to get a thousand witnesses simply to swear that they did not see him in such and such a place, unless the witnesses are prepared to prove that they must have seen him had he been there. But the evidence that animal life commenced with the Lingula-flags, *e.g.*, would seem to be exactly of this unsatisfactory uncorroborated sort. The Cambrian witnesses simply swear they " haven't seen anybody their way "; upon which the counsel for the other side immediately puts in ten or twelve thousand feet of Devonian sandstones to make oath they never saw a fish or a mollusk, though all the world knows there were plenty in their time.

But then it is urged that, though the Devonian rocks in one part of the world exhibit no fossils, in another they do, while the lower Cambrian rocks nowhere exhibit fossils, and hence no living being could have existed in their epoch.

To this there are two replies : the first that the observational basis of the assertion that the lowest rocks are nowhere fossiliferous is an amazingly small one, seeing how very small an area, in comparison to that of the whole world, has yet been fully searched ; the second, that the argument is good for nothing unless the unfossiliferous rocks in question were not only *contemporaneous* in the geological sense, but *synchronous* in the chronological sense. To use the *alibi* illustration again.

If a man wishes to prove he was in neither of two places, A and B, on a given day, his witnesses for each place must be prepared to answer for the whole day. If they can only prove that he was not at A in the morning, and not at B in the afternoon, the evidence of his absence from both is *nil*, because he might have been at B in the morning and at A in the afternoon.

Thus everything depends upon the validity of the second assumption. And we must proceed to inquire what is the real meaning of the word "contemporaneous" as employed by geologists. To this end a concrete example may be taken.

The Lias of England and the Lias of Germany, the Cretaceous rocks of Britain and the Cretaceous rocks of Southern India, are termed by geologists "contemporaneous" formations; but whenever any thoughtful geologist is asked whether he means to say that they were deposited synchronously, he says, "No,—only within the same great epoch." And if, in pursuing the inquiry, he is asked what may be the approximate value in time of a "great epoch"—whether it means a hundred years, or a thousand, or a million, or ten million years—his reply is, "I cannot tell."

If the further question be put, whether physical geology is in possession of any method by which the actual synchrony (or the reverse) of any two distant deposits can be ascertained, no such method can be heard of; it being admitted by all the best authorities that neither similarity of mineral composition, nor of physical character, nor even direct continuity of stratum, are *absolute* proofs of the synchronism of even approximated sedimentary strata : while, for distant deposits, there seems to be no kind of physical evidence attainable of a nature competent to decide whether such deposits were formed simultaneously, or whether they possess any given difference of antiquity. To return to an example already given : All competent authorities will probably assent to the proposition that physical geology does not enable us in any way to reply to this question—Were the British Cretaceous rocks deposited at the same time as those of India, or are they a million of years younger or a million of years older ?

Is palæontology able to succeed where physical geology fails ? Standard writers on palæontology, as has been seen, assume that she can. They take it for granted, that deposits containing similar organic remains are synchronous—at any rate in a broad sense ; and yet, those who will study the eleventh and twelfth chapters of Sir Henry De La Beche's remarkable " Researches in Theoretical Geology," published now nearly thirty years ago, and will carry out the arguments there most luminously stated, to their logical consequences, may very easily convince themselves that even absolute identity of organic contents is no proof of the synchrony of deposits, while absolute diversity is no proof of difference of date. Sir Henry De La Beche goes even further, and adduces conclusive evidence to show that the different parts of one and the same stratum, having a similar composition throughout, containing the same organic remains, and having similar beds above and below it, may yet differ to any conceivable extent in age.

Edward Forbes was in the habit of asserting that the similarity of the organic contents of distant formations was *primâ facie* evidence, not of their similarity, but of their difference of age ; and holding as he did the doctrine of single specific centres, the conclusion was as legitimate as any other ; for the two districts must have been occupied by migration from one of the two, or from an intermediate spot, and the chances against exact coincidence of migration and of imbedding are infinite.

In point of fact, however, whether the hypothesis of single or of multiple specific centres be adopted, similarity of organic contents cannot possibly afford any proof of the synchrony of the deposits which contain them ; on the contrary, it is demonstrably compatible with the lapse of the most prodigious intervals of time, and with the interposition of vast changes in the organic and inorganic worlds, between the epochs in which such deposits were formed.

On what amount of similarity of their faunæ is the doctrine of the contemporaneity of the European and of the North American Silurians based? In the last edition of Sir Charles Lyell's "Elementary Geology" it is stated, on the authority of a former President of this Society, the late Daniel Sharpe, that between 30 and 40 per cent. of the species of Silurian Mollusca are common to both sides of the Atlantic. By way of due allowance for further discovery, let us double the lesser number and suppose that 60 per cent. of the species are common to the North American and the British Silurians. Sixty per cent. of species in common is, then, proof of contemporaneity.

Now suppose that, a million or two of years hence, when Britain has made another dip beneath the sea and has come up again, some geologist applies this doctrine, in comparing the strata laid bare by the upheaval of the bottom, say, of St. George's Channel with what may then remain of the Suffolk Crag. Reasoning in the same way, he will at once decide the Suffolk Crag and the St. George's Channel beds to be contemporaneous; although we happen to know that a vast period (even in the geological sense) of time, and physical changes of almost unprecedented extent, separate the two.

But if it be a demonstrable fact that strata containing more than 60 or 70 per cent. of species of Mollusca in common, and comparatively close together, may yet be separated by an amount of geological time sufficient to allow of some of the greatest physical changes the world has seen, what becomes of that sort of contemporaneity the sole evidence of which is a similarity of facies, or the identity of half a dozen species, or of a good many genera?

And yet there is no better evidence for the contemporaneity assumed by all who adopt the hypothesis of universal faunæ and floræ, of a universally uniform climate, and of a sensible cooling of the globe during geological time.

There seems, then, no escape from the admission that neither physical geology, nor palæontology, possesses any method by which the absolute synchronism of two strata can be demonstrated. All that geology can prove is local order of succession. It is mathematically certain that, in any given vertical linear section of an undisturbed series of sedimentary deposits, the bed which lies lowest is the oldest. In many other vertical linear sections of the same series, of course, corresponding beds will occur in a similar order; but, however great may be the probability, no man can say with absolute certainty that the beds in the two sections were synchronously deposited. For areas of moderate extent, it is doubtless true that no practical evil is likely to result from assuming the corresponding beds to be synchronous or strictly contemporaneous; and there are multitudes of accessory circumstances which may fully justify the assumption of such synchrony But the moment the geologist has to deal with large areas, or with completely separated deposits, the mischief of confounding that " homotaxis " or " similarity of arrangement," which *can* be demonstrated, with " synchrony " or "identity of date," for which there is not a shadow of proof, under the one common term of " contemporaneity " becomes incalculable, and proves the constant source of gratuitous speculations.

For anything that geology or palæontology are able to show to the contrary, a Devonian fauna and flora in the British Islands may have been contemporaneous with Silurian life in North America, and with a Carboniferous fauna and flora in Africa. Geographical provinces and zones may have been as distinctly marked in the Palæozoic epoch as at present, and those seemingly sudden appearances of new genera and species, which we ascribe to new creation, may be simple results of migration.

It may be so; it may be otherwise. In the present condition of our knowledge and of our methods, one verdict—" not proven, and not provable "—must be recorded against all the grand hypotheses of the palæontologist respecting the general succession of life on the globe. The

order and nature of terrestrial life, as a whole,
are open questions. Geology at present provides
us with most valuable topographical records, but
she has not the means of working them into a
universal history. Is such a universal history,
then, to be regarded as unattainable? Are all
the grandest and most interesting problems which
offer themselves to the geological student, essentially insoluble? Is he in the position of a scientific
Tantalus—doomed always to thirst for a knowledge
which he cannot obtain? The reverse is to be
hoped; nay, it may not be impossible to indicate
the source whence help will come.

In commencing these remarks, mention was
made of the great obligations under which the
naturalist lies to the geologist and palæontologist.
Assuredly the time will come when these obligations will be repaid tenfold, and when the maze of
the world's past history, through which the pure
geologist and the pure palæontologist find no
guidance, will be securely threaded by the clue
furnished by the naturalist.

All who are competent to express an opinion on
the subject are, at present, agreed that the mani-
fold varieties of animal and vegetable form have
not either come into existence by chance, nor
result from capricious exertions of creative power;
but that they have taken place in a definite order,
the statement of which order is what men of
science term a natural law. Whether such a law

is to be regarded as an expression of the mode of operation of natural forces, or whether it is simply a statement of the manner in which a supernatural power has thought fit to act, is a secondary question, so long as the existence of the law and the possibility of its discovery by the human intellect are granted. But he must be a half-hearted philosopher who, believing in that possibility, and having watched the gigantic strides of the biological sciences during the last twenty years, doubts that science will sooner or later make this further step, so as to become possessed of the law of evolution of organic forms—of the unvarying order of that great chain of causes and effects of which all organic forms, ancient and modern, are the links. And then, if ever, we shall be able to begin to discuss, with profit, the questions respecting the commencement of life, and the nature of the successive populations of the globe, which so many seem to think are already answered.

The preceding arguments make no particular claim to novelty ; indeed they have been floating more or less distinctly before the minds of geologists for the last thirty years ; and if, at the present time, it has seemed desirable to give them more definite and systematic expression, it is because palæontology is every day assuming a greater importance, and now requires to rest on a basis the firmness of which is thoroughly well assured. Among its fundamental conceptions, there must be no confusion between what is certain and what is more or less probable.[1]   But, pending the construction of a surer foundation than palæontology now possesses, it may be instructive, assuming for the nonce the general correctness of the ordinary hypothesis of geological contemporaneity, to consider whether the deductions which are ordinarily drawn from the whole body of palæontological facts are justifiable.

The evidence on which such conclusions are based is of two kinds, negative and positive.   The value of negative evidence, in connection with this inquiry, has been so fully and clearly discussed in an address from the chair of this Society,[2] which none of us have forgotten, that nothing need at present be said about it; the more, as the considerations which have been laid before you have certainly not tended to increase your estimation of such evidence.   It will be preferable to turn to the positive facts of palæontology, and to inquire what they tell us.

We are all accustomed to speak of the number and the extent of the changes in the living population of the globe during geological time as

---

[1] " Le plus grand service qu'on puisse rendre à la science est d'y faire place nette avant d'y rien construire."—CUVIER.
[2] Anniversary Address for 1851, *Quart. Journ. Geol. Soc.* vol. vii.

something enormous : and indeed they are so, if
we regard only the negative differences which
separate the older rocks from the more modern,
and if we look upon specific and generic changes
as great changes, which from one point of view,
they truly are. But leaving the negative differences out of consideration, and looking only at the
positive data furnished by the fossil world from a
broader point of view—from that of the comparative anatomist who has made the study of the
greater modifications of animal form his chief
business—a surprise of another kind dawns upon
the mind ; and under *this* aspect the smallness of
the total change becomes as astonishing as was its
greatness under the other.

There are two hundred known orders of plants ;
of these not one is certainly known to exist exclusively in the fossil state. The whole lapse of
geological time has as yet yielded not a single new
ordinal type of vegetable structure.[1]

The positive change in passing from the recent
to the ancient animal world is greater, but still
singularly small. No fossil animal is so distinct
from those now living as to require to be arranged
even in a separate class from those which contain
existing forms. It is only when we come to the
orders, which may be roughly estimated at about
a hundred and thirty, that we meet with fossil

---

[1] See Hooker's *Introductory Essay to the Flora of Tasmania,*
p. xxiii.

animals so distinct from those now living as to require orders for themselves; and these do not amount, on the most liberal estimate, to more than about 10 per cent. of the whole.

There is no certainly known extinct order of Protozoa; there is but one among the Cœlenterata —that of the rugose corals; there is none among the Mollusca; there are three, the Cystidea, Blastoidea, and Edrioasterida, among the Echinoderms; and two, the Trilobita and Eurypterida, among the Crustacea; making altogether five for the great sub-kingdom of Annulosa. Among Vertebrates there is no ordinally distinct fossil fish : there is only one extinct order of Amphibia —the Labyrinthodonts; but there are at least four distinct orders of Reptilia, viz. the Ichthyosauria, Plesiosauria, Pterosauria, Dinosauria, and perhaps another or two. There is no known extinct order of Birds, and no certainly known extinct order of Mammals, the ordinal distinctness of the " Toxodontia " being doubtful.

The objection that broad statements of this kind, after all, rest largely on negative evidence is obvious, but it has less force than may at first be supposed; for, as might be expected from the circumstances of the case, we possess more abundant positive evidence regarding Fishes and marine Mollusks than respecting any other forms of animal life; and yet these offer us, through the whole range of geological time, no species ordinally

distinct from those now living; while the far less
numerous class of Echinoderms presents three, and
the Crustacea two, such orders, though none of these
come down later than the Palæozoic age. Lastly,
the Reptilia present the extraordinary and exceptional phenomenon of as many extinct as existing
orders, if not more; the four mentioned maintaining their existence from the Lias to the Chalk
inclusive.

Some years ago one of your Secretaries pointed
out another kind of positive palæontological
evidence tending towards the same conclusion—
afforded by the existence of what he termed
"persistent types" of vegetable and of animal
life.[1] He stated, on the authority of Dr. Hooker,
that there are Carboniferous plants which appear
to be generically identical with some now living;
that the cone of the Oolitic *Araucaria* is hardly
distinguishable from that of an existing species;
that a true *Pinus* appears in the Purbecks and a
Juglans in the Chalk; while, from the Bagshot
Sands, a *Banksia*, the wood of which is not
distinguishable from that of species now living
in Australia, had been obtained.

Turning to the animal kingdom, he affirmed
the tabulate corals of the Silurian rocks to be
wonderfully like those which now exist; while

[1] See the abstract of a Lecture "On the Persistent Types of
Animal Life," in the *Notices of the Meetings of the Royal
Institution of Great Britain.*—June 3, 1859, vol. iii. p. 151.

even the families of the Aporosa were all represented in the older Mesozoic rocks.

Among the Mollusca similar facts were adduced.
Let it be borne in mind that *Avicula, Mytilus,
Chiton, Natica, Patella, Trochus, Discina, Orbicula,
Lingula, Rhynchonella,* and *Nautilus,* all of which
are existing *genera,* are given without a doubt as
Silurian in the last edition of "Siluria"; while
the highest forms of the highest Cephalopods
are represented in the Lias by a genus *Belemno-
teuthis,* which presents the closest relation to the
existing *Loligo.*

The two highest groups of the Annulosa, the
Insecta and the Arachnida, are represented in the
Coal, either by existing genera, or by forms
differing from existing genera in quite minor
peculiarities.

Turning to the Vertebrata, the only palæozoic
Elasmobranch Fish of which we have any complete
knowledge is the Devonian and Carboniferous
Pleuracanthus, which differs no more from existing
Sharks than these do from one another.

Again, vast as is the number of undoubtedly
Ganoid fossil Fishes, and great as is their range
in time, a large mass of evidence has recently
been adduced to show that almost all those
respecting which we possess sufficient information,
are referable to the same sub-ordinal groups as
the existing *Lepidosteus, Polypterus,* and Sturgeon;
and that a singular relation obtains between the

older and the younger Fishes; the former, the
Devonian Ganoids, being almost all members of
the same sub-order as *Polypterus*, while the
Mesozoic Ganoids are almost all similarly allied
to *Lepidosteus*.[1]

Again, what can be more remarkable than the
singular constancy of structure preserved throughout a vast period of time by the family of the
Pycnodonts and by that of the true Cœlacanths:
the former persisting, with but insignificant
modifications, from the Carboniferous to the
Tertiary rocks, inclusive; the latter existing,
with still less change, from the Carboniferous
rocks to the Chalk, inclusive?

Among Reptiles, the highest living group, that
of the Crocodilia, is represented, at the early part
of the Mesozoic epoch, by species identical in the
essential characters of their organisation with
those now living, and differing from the latter
only in such matters as the form of the articular
facets of the vertebral centra, in the extent to
which the nasal passages are separated from the
cavity of the mouth by bone, and in the proportions of the limbs.

And even as regards the Mammalia, the scanty
remains of Triassic and Oolitic species afford no
foundation for the supposition that the organisa-

---

[1] "Memoirs of the Geological Survey of the United Kingdom.
—Decade x. Preliminary Essay upon the Systematic Arrangement of the Fishes of the Devonian Epoch."

tion of the oldest forms differed nearly so much from some of those which now live as these differ from one another.

It is needless to multiply these instances; enough has been said to justify the statement that, in view of the immense diversity of known animal and vegetable forms, and the enormous lapse of time indicated by the accumulation of fossiliferous strata, the only circumstance to be wondered at is, not that the changes of life, as exhibited by positive evidence, have been so great, but that they have been so small.

Be they great or small, however, it is desirable to attempt to estimate them. Let us, therefore, take each great division of the animal world in succession, and, whenever an order or a family can be shown to have had a prolonged existence, let us endeavour to ascertain how far the later members of the group differ from the earlier ones. If these later members, in all or in many cases, exhibit a certain amount of modification, the fact is, so far, evidence in favour of a general law of change; and, in a rough way, the rapidity of that change will be measured by the demonstrable amount of modification. On the other hand, it must be recollected that the absence of any modification, while it may leave the doctrine of the existence of a law of change without positive support, cannot possibly disprove all forms of that doctrine, though

it may afford a sufficient refutation of many of
them.

The PROTOZOA.—The Protozoa are represented
throughout the whole range of geological series,
from the Lower Silurian formation to the present
day. The most ancient forms recently made
known by Ehrenberg are exceedingly like those
which now exist : no one has ever pretended that
the difference between any ancient and any modern
Foraminifera is of more than generic value, nor
are the oldest Foraminifera either simpler, more
embryonic, or less differentiated, than the existing
forms.

The CŒLENTERATA.—The Tabulate Corals have
existed from the Silurian epoch to the present
day, but I am not aware that the ancient *Heliolites*
possesses a single mark of a more embryonic or
less differentiated character, or less high organisation, than the existing *Heliopora*. As for the
Aporose Corals, in what respect is the Silurian
Palæocyclus less highly organised or more embryonic than the modern *Fungia*, or the Liassic
Aporosa than the existing members of the same
families ?

The *Mollusca*.—In what sense is the living
Waldheimia less embryonic, or more specialised,
than the palæozoic *Spirifer ;* or the existing
Rhynchonellæ, Craniæ, Discinæ, Lingulæ, than the
Silurian species of the same genera ? In what
sense can *Loligo* or *Spirula* be said to be more

specialised, or less embryonic, than *Belemnites*; or the modern species of Lamellibranch and Gasteropod genera, than the Silurian species of the same genera?

The ANNULOSA.—The Carboniferous Insecta and Arachnida are neither less specialised, nor more embryonic, than these that now live, nor are the Liassic Cirripedia and Macrura; while several of the Brachyura, which appear in the Chalk, belong to existing genera; and none exhibit either an intermediate, or an embryonic, character.

The VERTEBRATA.—Among fishes I have referred to the Cœlacanthini (comprising the genera *Cœlacanthus, Holophagus, Undina,* and *Macropoma*) as affording an example of a persistent type; and it is most remarkable to note the smallness of the differences between any of these fishes (affecting at most the proportions of the body and fins, and the character and sculpture of the scales), notwithstanding their enormous range in time. In all the essentials of its very peculiar structure, the *Macropoma* of the Chalk is identical with the *Cœlacanthus* of the Coal. Look at the genus *Lepidotus*, again, persisting without a modification of importance from the Liassic to the Eocene formations inclusivly.

Or among the Teleostei—in what respect is the *Beryx* of the Chalk more embryonic, or

less differentiated, than *Beryx lineatus* of King
George's Sound ?

Or to turn to the higher Vertebrata—in what
sense are the Liassic Chelonia inferior to those
which now exist? How are the Cretaceous
Ichthyosauria, Plesiosauria, or Pterosauria less
embryonic, or more differentiated, species than
those of the Lias ?

Or lastly, in what circumstance is the *Phascolotherium* more embryonic, or of a more generalised type, than the modern Opossum; or a
Lophiodon, or a *Palœotherium*, than a modern
Tapirus or *Hyrax?*

These examples might be almost indefinitely
multiplied, but surely they are sufficient to prove
that the only safe and unquestionable testimony
we can procure—positive evidence—fails to demonstrate any sort of progressive modification
towards a less embryonic, or less generalised, type
in a great many groups of animals of long-
continued geological existence. In these groups
there is abundant evidence of variation—none of
what is ordinarily understood as progression; and,
if the known geological record is to be regarded
as even any considerable fragment of the whole,
it is inconceivable that any theory of a necessarily
progressive development can stand, for the numerous orders and families cited afford no trace of
such a process.

But it is a most remarkable fact, that, while the

groups which have been mentioned, and many besides, exhibit no sign of progressive modification, there are others, co-existing with them, under the same conditions, in which more or less distinct indications of such a process seems to be traceable. Among such indications I may remind you of the predominance of Holostome Gasteropoda in the older rocks as compared with that of Siphonostone Gasteropoda in the later. A case less open to the objection of negative evidence, however, is that afforded by the Tetrabranchiate Cephalopoda, the forms of the shells and of the septal sutures exhibiting a certain increase of complexity in the newer genera. Here, however, one is met at once with the occurrence of *Orthoceras* and *Baculites* at the two ends of the series, and of the fact that one of the simplest genera, *Nautilus*, is that which now exists.

The Crinoidea, in the abundance of stalked forms in the ancient formations as compared with their present rarity, seem to present us with a fair case of modification from a more embryonic towards a less embryonic condition. But then, on careful consideration of the facts, the objection arises that the stalk, calyx, and arms of the palæozoic Crinoid are exceedingly different from the corresponding organs of a larval *Comatula ;* and it might with perfect justice be argued that *Actinocrinus* and *Eucalyptocrinus*, for example, depart to the full as widely, in one direction, from the stalked embryo of *Comatula*, as *Comatula* itself does in the other.

The Echinidea, again, are frequently quoted as exhibiting a gradual passage from a more generalised to a more specialised type, seeing that the elongated, or oval, Spatangoids appear after the spheroidal Echinoids. But here it might be argued, on the other hand, that the spheroidal Echinoids, in reality, depart further from the general plan and from the embryonic form than the elongated Spatangoids do; and that the peculiar dental apparatus and the pedicellariæ of the former are marks of at least as great differentiation as the petaloid ambulacra and semitæ of the latter.

Once more, the prevalence of Macrurous before Brachyurous Podophthalmia is, apparently, a fair piece of evidence in favour of progressive modification in the same order of Crustacea; and yet the case will not stand much sifting, seeing that the Macrurous Podophthalmia depart as far in one direction from the common type of Podophthalmia, or from any embryonic condition of the Brachyura, as the Brachyura do in the other; and that the middle terms between Macrura and Brachyura—the Anomura—are little better represented in the older Mesozoic rocks than the Brachyura are.

None of the cases of progressive modification which are cited from among the Invertebrata appear to me to have a foundation less open to

criticism than these ; and if this be so, no careful
reasoner would, I think, be inclined to lay very
great stress upon them. Among the Vertebrata,
however, there are a few examples which appear to
be far less open to objection.

It is, in fact, true of several groups of Vertebrata which have lived through a considerable
range of time, that the endoskeleton (more particularly the spinal column) of the older genera
presents a less ossified, and, so far, less differentiated, condition than that of the younger
genera. Thus the Devonian Ganoids, though
almost all members of the same sub-order as
Polypterus, and presenting numerous important resemblances to the existing genus, which possesses
biconclave vertebræ, are, for the most part, wholly
devoid of ossified vertebral centra. The Mesozoic
Lepidosteidæ, again, have, at most, biconcave
vertebræ, while the existing *Lepidosteus* has
Salamandroid, opisthocœlous, vertebræ. So, none
of the Palæozoic Sharks have shown themselves
to be possessed of ossified vertebræ, while the
majority of modern Sharks possess such vertebræ.
Again, the more ancient Crocodilia and Lacertilia
have vertebræ with the articular facets of their
centra flattened or biconcave, while the modern
members of the same group have them procœlous.
But the most remarkable examples of progressive
modification of the vertebral column, in correspondence with geological age, are those afforded

by the Pycnodonts among fish, and the Labyrinthodonts among Amphibia.

The late able ichthyologist Heckel pointed out the fact, that, while the Pycnodonts never possess true vertebral centra, they differ in the degree of expansion and extension of the ends of the bony arches of the vertebræ upon the sheath of the notochord ; the Carboniferous forms exhibiting hardly any such expansion, while the Mesozoic genera present a greater and greater development, until, in the Tertiary forms, the expanded ends become suturally united so as to form a sort of false vertebra. Hermann von Meyer, again, to whose luminous researches we are indebted for our present large knowledge of the organisation of the older Labyrinthodonts, has proved that the Carboniferous *Archegosaurus* had very imperfectly developed vertebral centra, while the Triassic *Mastodonsaurus* had the same parts completely ossified.[1]

The regularity and evenness of the dentition of the *Anoplotherium*, as contrasted with that of existing Artiodactyles, and the assumed nearer approach of the dentition of certain ancient Carnivores to the typical arrangement, have also been cited as exemplifications of a law of progressive development, but I know of no other cases based on

[1] As this Address is passing through the press (March 7, 1862), evidence lies before me of the existence of a new Labyrinthodont (*Pholidogaster*), from the Edinburgh coal-field with well-ossified vertebral centra.

positive evidence which are worthy of particular notice.

What then does an impartial survey of the positively ascertained truths of palæontology testify in relation to the common doctrines of progressive modification, which suppose that modification to have taken place by a necessary progress from more to less embryonic forms, or from more to less generalised types, within the limits of the period represented by the fossiliferous rocks?

It negatives those doctrines; for it either shows us no evidence of any such modification, or demonstrates it to have been very slight; and as to the nature of that modification, it yields no evidence whatsoever that the earlier members of any long-continued group were more generalised in structure than the later ones. To a certain extent, indeed, it may be said that imperfect ossification of the vertebral column is an embryonic character; but, on the other hand, it would be extremely incorrect to suppose that the vertebral columns of the older Vertebrata are in any sense embryonic in their whole structure.

Obviously, if the earliest fossiliferous rocks now known are coëval with the commencement of life, and if their contents give us any just conception of the nature and the extent of the earliest fauna and flora, the insignificant amount of modification which can be demonstrated to have taken place in any one group of animals, or plants, is quite incompatible with the hypothesis that all living forms are the results of a necessary process of progressive development, entirely comprised within the time represented by the fossiliferous rocks.

Contrariwise, any admissible hypothesis of progressive modification must be compatible with persistence without progression, through indefinite periods. And should such an hypothesis eventually be proved to be true, in the only way in which it can be demonstrated, viz. by observation and experiment upon the existing forms of life, the conclusion will inevitably present itself, that the Palæozoic Mesozoic, and Cainozoic faunæ and floræ, taken together, bear somewhat the same proportion to the whole series of living beings which have occupied this globe, as the existing fauna and flora do to them.

Such are the results of palæontology as they appear, and have for some years appeared, to the mind of an inquirer who regards that study simply as one of the applications of the great biological sciences, and who desires to see it placed upon the same sound basis as other branches of physical inquiry. If the arguments which have been brought forward are valid, probably no one, in view of the present state of opinion, will be inclined to think the time wasted which has been spent upon their elaboration.

X

GEOLOGICAL REFORM

[1869]

"A great reform in geological speculation seems now to have become necessary."

"It is quite certain that a great mistake has been made—that British popular geology at the present time is in direct opposition to the principles of Natural Philosophy." [1]

IN reviewing the course of geological thought during the past year, for the purpose of discovering those matters to which I might most fitly direct your attention in the Address which it now becomes my duty to deliver from the Presidential Chair, the two somewhat alarming sentences which I have just read, and which occur in an able and interesting essay by an eminent natural philosopher, rose into such prominence before my mind that they eclipsed everything else.

It surely is a matter of paramount importance

---

[1] On Geological Time. By Sir W. Thomson, LL.D. *Transactions of the Geological Society of Glasgow*, vol. iii.

for the British geologists (some of them very popular geologists too) here in solemn annual session assembled, to inquire whether the severe judgment thus passed upon them by so high an authority as Sir William Thomson is one to which they must plead guilty *sans phrase*, or whether they are prepared to say "not guilty," and appeal for a reversal of the sentence to that higher court of educated scientific opinion to which we are all amenable.

As your attorney-general for the time being, I thought I could not do better than get up the case with a view of advising you. It is true that the charges brought forward by the other side involve the consideration of matters quite foreign to the pursuits with which I am ordinarily occupied ; but, in that respect, I am only in the position which is, nine times out of ten, occupied by counsel, who nevertheless contrive to gain their causes, mainly by force of mother-wit and common-sense, aided by some training in other intellectual exercises.

Nerved by such precedents, I proceed to put my pleading before you.

And the first question with which I propose to deal is, What is it to which Sir W. Thomson refers when he speaks of "geological speculation" and "British popular geology"?

I find three, more or less contradictory, systems of geological thought, each of which might fairly enough claim these appellations, standing side by side in Britain. I shall call one of them CATA-STROPHISM, another UNIFORMITARIANISM, the third EVOLUTIONISM ; and I shall try briefly to sketch the characters of each, that you may say whether the classification is, or is not, exhaustive.

By CATASTROPHISM, I mean any form of geological speculation which, in order to account for the phenomena of geology, supposes the operation of forces different in their nature, or immeasurably different in power, from those which we at present see in action in the universe.

The Mosaic cosmogony is, in this sense, catastrophic, because it assumes the operation of extra-natural power. The doctrine of violent upheavals, *débâcles*, and cataclysms in general, is catastrophic, so far as it assumes that these were brought about by causes which have now no parallel. There was a time when catastrophism might, pre-eminently, have claimed the title of " British popular geology " ; and assuredly it has yet many adherents, and reckons among its supporters some of the most honoured members of this Society.

By UNIFORMITARIANISM, I mean especially, the teaching of Hutton and of Lyell.

That great though incomplete work, " The Theory of the Earth," seems to me to be one of the most remarkable contributions to geology which is recorded in the annals of the science.

So far as the not-living world is concerned, uniformitarianism lies there, not only in germ, but in blossom and fruit.

If one asks how it is that Hutton was led to entertain views so far in advance of those prevalent in his time, in some respects ; while, in others, they seem almost curiously limited, the answer appears to me to be plain.

Hutton was in advance of the geological speculation of his time, because, in the first place, he had amassed a vast store of knowledge of the facts of geology, gathered by personal observation in travels of considerable extent ; and because, in the second place, he was thoroughly trained in the physical and chemical science of his day, and thus possessed, as much as any one in his time could possess it, the knowledge which is requisite for the just interpretation of geological phenomena, and the habit of thought which fits a man for scientific inquiry.

It is to this thorough scientific training that I ascribe Hutton's steady and persistent refusal to look to other causes than those now in operation, for the explanation of geological phenomena.

Thus he writes :—" I do not pretend, as he [M. de Luc] does in his theory, to describe the beginning of things. I take things such as I find them at present ; and from these I reason with regard to that which must have been." [1]

----

[1] *The Theory of the Earth*, vol. i. p. 173, note.

And again :—" A theory of the earth, which has for object truth, can have no retrospect to that which had preceded the present order of the world; for this order alone is what we have to reason upon; and to reason without data is nothing but delusion. A theory, therefore, which is limited to the actual constitution of this earth cannot be allowed to proceed one step beyond the present order of things." [1]

And so clear is he, that no causes beside such as are now in operation are needed to account for the character and disposition of the components of the crust of the earth, that he says, broadly and boldly :—" . . . There is no part of the earth which has not had the same origin, so far as this consists in that earth being collected at the bottom of the sea, and afterwards produced, as land, along with masses of melted substances, by the operation of mineral causes." [2]

But other influences were at work upon Hutton beside those of a mind logical by nature, and scientific by sound training; and the peculiar turn which his speculations took seems to me to be unintelligible, unless these be taken into account. The arguments of the French astronomers and mathematicians, which, at the end of the last century, were held to demonstrate the existence of a compensating arrangement among the celestial bodies, whereby all perturba-

[1] *The Theory of the Earth*, vol. i. p. 281.   [2] *Ibid.* p. 371.

tions eventually reduced themselves to oscillations on each side of a mean position, and the
stability of the solar system was secured, had
evidently taken strong hold of Hutton's mind.

In those oddly constructed periods which seem
to have prejudiced many persons against reading
his works, but which are full of that peculiar, if
unattractive, eloquence which flows from mastery
of the subject, Hutton says:—

"We have now got to the end of our reasoning;
we have no data further to conclude immediately
from that which actually is. But we have got
enough; we have the satisfaction to find, that
in Nature there is wisdom, system, and consistency. For having, in the natural history of this
earth, seen a succession of worlds, we may from
this conclude that there is a system in Nature;
in like manner as, from seeing revolutions of the
planets, it is concluded, that there is a system by
which they are intended to continue those revolutions. But if the succession of worlds is established in the system of nature, it is in vain to
look for anything higher in the origin of the
earth. The result, therefore, of this physical
inquiry is, that we find no vestige of a beginning,
—no prospect of an end." [1]

Yet another influence worked strongly upon
Hutton. Like most philosophers of his age, he
coquetted with those final causes which have

[1] *The Theory of the Earth*, vol. i. p. 200.

been named barren virgins, but which might
be more fitly termed the *hetairæ* of philosophy,
so constantly have they led men astray. The
final cause of the existence of the world is, for
Hutton, the production of life and intelligence.

"We have now considered the globe of this
earth as a machine, constructed upon chemical
as well as mechanical principles, by which its
different parts are all adapted, in form, in quality,
and in quantity, to a certain end; an end attained with certainty or success; and an end
from which we may perceive wisdom, in contem-
plating the means employed.

"But is this world to be considered thus
merely as a machine, to last no longer than its
parts retain their present position, their proper
forms and qualities? Or may it not be also
considered as an organised body? such as has
a constitution in which the necessary decay of
the machine is naturally repaired, in the exertion
of those productive powers by which it had
been formed.

"This is the view in which we are now to
examine the globe; to see if there be, in the
constitution of this world, a reproductive operation, by which a ruined constitution may be
again repaired, and a duration or stability thus
procured to the machine, considered as a world
sustaining plants and animals."[1]

[1] *The Theory of the Earth*, vol. i. pp. 16, 17.

Kirwan, and the other Philistines of the day, accused Hutton of declaring that his theory implied that the world never had a beginning, and never differed in condition from its present state. Nothing could be more grossly unjust, as he expressly guards himself against any such conclusion in the following terms :—

" But in thus tracing back the natural operations which have succeeded each other, and mark to us the course of time past, we come to a period in which we cannot see any farther. This, however, is not the beginning of the operations which proceed in time and according to the wise economy of this world ; nor is it the establishing of that which, in the course of time, had no beginning ; it is only the limit of our retrospective view of those operations which have come to pass in time, and have been conducted by supreme intelligence." [1]

I have spoken of Uniformitarianism as the doctrine of Hutton and of Lyell. If I have quoted the older writer rather than the newer, it is because his works are little known, and his claims on our veneration too frequently forgotten, not because I desire to dim the fame of his eminent successor. Few of the present generation of geologists have read Playfair's " Illustrations," fewer still the original " Theory of the Earth " ; the more is the pity ; but which of us has not thumbed

---

[1] *The Theory of the Earth,* vol. i. p. 223.

every page of the " Principles of Geology " ? I
think that he who writes fairly the history of his
own progress in geological thought, will not be
able to separate his debt to Hutton from his
obligations to Lyell ; and the history of the progress of individual geologists is the history of
geology.

No one can doubt that the influence of uniformitarian views has been enormous, and, in the
main, most beneficial and favourable to the
progress of sound geology.

Nor can it be questioned that Uniformitarianism
has even a stronger title than Catastrophism to
call itself the geological speculation of Britain, or,
if you will, British popular geology. For it is
eminently a British doctrine, and has even now
made comparatively little progress on the continent of Europe. Nevertheless, it seems to me
to be open to serious criticism upon one of its
aspects.

I have shown how unjust was the insinuation
that Hutton denied a beginning to the world.
But it would not be unjust to say that he persistently in practice, shut his eyes to the existence
of that prior and different state of things which,
in theory, he admitted ; and, in this aversion to
look beyond the veil of stratified rocks, Lyell
follows him.

Hutton and Lyell alike agree in their indis-
position to carry their speculations a step beyond

the period recorded in the most ancient strata now open to observation in the crust of the earth. This is, for Hutton, " the point in which we cannot see any farther " ; while Lyell tells us,—

" The astronomer may find good reasons for ascribing the earth's form to the original fluidity of the mass, in times long antecedent to the first introduction of living beings into the planet ; but the geologist must be content to regard the earliest monuments which it is his task to interpret, as belonging to a period when the crust had already acquired great solidity and thickness, probably as great as it now possesses, and when volcanic rocks, not essentially differing from those now produced, were formed from time to time, the intensity of volcanic heat being neither greater nor less than it is now." [1]

And again, " As geologists, we learn that it is not only the present condition of the globe which has been suited to the accommodation of myriads of living creatures, but that many former states also have been adapted to the organisation and habits of prior races of beings. The disposition of the seas, continents and islands, and the climates, have varied ; the species likewise have been changed ; and yet they have all been so modelled, on types analogous to those of existing plants and animals, as to indicate, throughout, a perfect harmony of design and unity of purpose.   To

[1] *Principles of Geology*, vol. ii. p. 211.

assume that the evidence of the beginning, or end, of so vast a scheme lies within the reach of our philosophical inquiries, or even of our speculations, appears to be inconsistent with a just estimate of the relations which subsist between the finite powers of man and the attributes of an infinite and eternal Being." [1]

The limitations implied in these passages appear to me to constitute the weakness and the logical defect of Uniformitarianism. No one will impute blame to Hutton that, in face of the imperfect condition, in his day, of those physical sciences which furnish the keys to the riddles of geology, he should have thought it practical wisdom to limit his theory to an attempt to account for " the present order of things"; but I am at a loss to comprehend why, for all time, the geologist must be content to regard the oldest fossiliferous rocks as the *ultima Thule* of his science ; or what there is inconsistent with the relations between the finite and the infinite mind, in the assumption, that we may discern somewhat of the beginning, or of the end, of this speck in space we call our earth. The finite mind is certainly competent to trace out the development of the fowl within the egg ; and I know not on what ground it should find more difficulty in unravelling the complexities of the development of the earth. In fact, as Kant

---

[1] *Principles of Geology*, vol. ii. p. 613.

has well remarked,[1] the cosmical process is really simpler than the biological.

This attempt to limit, at a particular point, the progress of inductive and deductive reasoning from the things which are, to those which were—this faithlessness to its own logic, seems to me to have cost Uniformitarianism the place, as the permanent form of geological speculation, which it might otherwise have held.

It remains that I should put before you what I understand to be the third phase of geological speculation—namely, EVOLUTIONISM.

I shall not make what I have to say on this head clear, unless I diverge, or seem to diverge, for a while, from the direct path of my discourse, so far as to explain what I take to be the scope of geology itself. I conceive geology to be the history of the earth, in precisely the same sense as biology is the history of living beings; and I trust you will not think that I am overpowered by the influence of a dominant pursuit if I say that I trace a close analogy between these two histories.

If I study a living being, under what heads does the knowledge I obtain fall? I can learn its structure, or what we call its ANATOMY; and its

---

[1] "Man darf es sich also nicht befremden lassen, wenn ich mich unterstehe zu sagen, dass eher die Bildung aller Himmels-körper, die Ursache ihrer Bewegungen, kurz der Ursprung der ganzen gegenwärtigen Verfassung des Weltbaues werden können eingesehen werden, ehe die Erzeugung eines einzigen Krautes oder einer Raupe aus mechanischen Gründen, deutlich und vollständig kund werden wird."—KANT's *Sämmtliche Werke*, Bd. i. p. 220.

DEVELOPMENT, or the series of changes which it
passes through to acquire its complete structure.
Then I find that the living being has certain
powers resulting from its own activities, and the
interaction of these with the activities of other
things—the knowledge of which is PHYSIOLOGY.
Beyond this the living being has a position in
space and time, which is its DISTRIBUTION. All
these form the body of ascertainable facts which
constitute the *status quo* of the living creature.
But these facts have their causes; and the
ascertainment of these causes is the doctrine of
ÆTIOLOGY.

If we consider what is knowable about the
earth, we shall find that such earth-knowledge—
if I may so translate the word geology—falls into
the same categories.

What is termed stratigraphical geology is neither
more nor less than the anatomy of the earth; and
the history of the succession of the formations is
the history of a succession of such anatomies, or
corresponds with development, as distinct from
generation.

The internal heat of the earth, the elevation
and depression of its crust, its belchings forth
of vapours, ashes, and lava, are its activities, in as
strict a sense as are warmth and the movements
and products of respiration the activities of an
animal. The phenomena of the seasons, of the
trade winds, of the Gulf-stream, are as much the

results of the reaction between these inner
activities and outward forces, as are the budding
of the leaves in spring and their falling in autumn
the effects of the interaction between the organisation of a plant and the solar light and heat.
And, as the study of the activities of the living being
is called its physiology, so are these phenomena
the subject-matter of an analogous telluric physiology, to which we sometimes give the name of
meteorology, sometimes that of physical geography,
sometimes that of geology.   Again, the earth has
a place in space and in time, and relations to
other bodies in both these respects, which constitute its distribution.   This subject is usually
left to the astronomer; but a knowledge of its
broad outlines seems to me to be an essential
constituent of the stock of geological ideas.

All that can be ascertained concerning the
structure, succession of conditions, actions, and position in space of the earth, is the matter of fact of its
natural history.   But, as in biology, there remains
the matter of reasoning from these facts to their
causes, which is just as much science as the other,
and indeed more; and this constitutes geological
ætiology.

' Having regard to this general scheme of geological knowledge and thought, it is obvious that
geological speculation may be, so to speak, ana-
tomical and developmental speculation, so far as it
relates to points of stratigraphical arrangement

which are out of reach of direct observation ; or, it may be physiological speculation so far as it relates to undetermined problems relative to the activities of the earth ; or, it may be distributional speculation, if it deals with modifications of the earth's place in space ; or, finally, it will be ætiological speculation if it attempts to deduce the history of the world, as a whole, from the known properties of the matter of the earth, in the conditions in which the earth has been placed.

For the purposes of the present discourse I may take this last to be what is meant by "geological speculation."

Now Uniformitarianism, as we have seen, tends to ignore geological speculation in this sense altogether.

The one point the catastrophists and the uniformitarians agreed upon, when this Society was founded, was to ignore it. And you will find, if you look back into our records, that our revered fathers in geology plumed themselves a good deal upon the practical sense and wisdom of this proceeding. As a temporary measure, I do not presume to challenge its wisdom ; but in all organised bodies temporary changes are apt to produce permanent effects ; and as time has slipped by, altering all the conditions which may have made such mortification of the scientific flesh desirable, I think the effect of the stream of cold water which has steadily flowed over geological speculation within these walls has been of doubtful beneficence.

The sort of geological speculation to which I am now referring (geological ætiology, in short) was created, as a science, by that famous philosopher Immanuel Kant, when, in 1775, he wrote his "General Natural History and Theory of the Celestial Bodies; or an Attempt to account for the Constitutional and the Mechanical Origin of the Universe upon Newtonian principles." [1]

In this very remarkable but seemingly little-known treatise,[2] Kant expounds a complete cosmogony, in the shape of a theory of the causes which have led to the development of the universe from diffused atoms of matter endowed with simple attractive and repulsive forces.

"Give me matter," says Kant, "and I will build the world;" and he proceeds to deduce from the simple data from which he starts, a doctrine in all essential respects similar to the well-known "Nebular Hypothesis" of Laplace.[1] He accounts for the relation of the masses and the densities of the planets to their distances from the sun, for the eccentricities of their orbits, for their rotations, for

---

[1] Grant (*History of Physical Astronomy*, p. 574) makes but the briefest reference to Kant.

[2] "Allgemeine Naturgeschichte und Theorie des Himmels; oder Versuch von der Verfassung und dem mechanischen Ursprunge des ganzen Weltgebäudes nach Newton'schen Grundsatzen abgehandelt."—KANT's *Sämmtliche Werke*, Bd. i. p. 207.   [3] *Système du Monde*, tome ii. chap. 6.

their satellites, for the general agreement in the direction of rotation among the celestial bodies, for Saturn's ring, and for the zodiacal light. He finds in each system of worlds, indications that the attractive force of the central mass will eventually destroy its organisation, by concentrating upon itself the matter of the whole system ; but, as the result of this concentration, he argues for the development of an amount of heat which will dissipate the mass once more into a molecular chaos such as that in which it began.

Kant pictures to himself the universe as once an infinite expansion of formless and diffused matter. At one point of this he supposes a single centre of attraction set up; and, by strict deductions from admitted dynamical principles, shows how this must result in the development of a prodigious central body, surrounded by systems of solar and planetary worlds in all stages of development. In vivid language he depicts the great world-maelstrom, widening the margins of its prodigious eddy in the slow progress of millions of ages, gradually reclaiming more and more of the molecular waste, and converting chaos into cosmos. But what is gained at the margin is lost in the centre; the attractions of the central systems bring their constituents together, which then, by the heat evolved, are converted once more into molecular chaos. Thus the worlds that are, lie between the ruins of the worlds that have been, and the chaotic materials of the worlds that shall be; and in spite of all waste and destruction, Cosmos is extending his borders at the expense of Chaos.

Kant's further application of his views to the earth itself is to be found in his " Treatise on Physical Geography "[1] (a term under which the then unknown science of geology was included), a subject which he had studied with very great care and on which he lectured for many years. The fourth section of the first part of this Treatise is called " History of the great Changes which the Earth has formerly undergone and is still undergoing," and is, in fact, a brief and pregnant essay upon the principles of geology. Kant gives an account first " of the gradual changes which are now taking place " under the heads of such as are caused by earthquakes, such as are brought about by rain and rivers, such as are effected by the sea, such as are produced by winds and frost; and, finally, such as result from the operations of man.

The second part is devoted to the " Memorials of the Changes which the Earth has undergone in remote Antiquity."   These are enumerated as :—
A. Proofs that the sea formerly covered the whole earth.   B. Proofs that the sea has often been changed into dry land and then again into sea. C. A discussion of the various theories of the

[1] KANT'S *Sämmtliche Werke*, Bd. viii. p. 145.

earth put forward by Scheuchzer, Moro, Bonnet,
Woodward, White, Leibnitz, Linnæus, and Buffon.

The third part contains an "Attempt to give a
sound explanation of the ancient history of the
earth."

I suppose that it would be very easy to pick
holes in the details of Kant's speculations, whether
cosmological, or specially telluric, in their application. But for all that, he seems to me to have
been the first person to frame a complete system
of geological speculation by founding the doctrine
of evolution.

With as much truth as Hutton, Kant could say,
" I take things just as I find them at present, and,
from these, I reason with regard to that which
must have been." Like Hutton, he is never tired
of pointing out that "in Nature there is wisdom,
system, and consistency." And, as in these great
principles, so in believing that the cosmos has a
reproductive operation " by which a ruined constitution may be repaired," he forestalls Hutton;
while, on the other hand, Kant is true to science.
He knows no bounds to geological speculation
but those of the intellect. He reasons back to a
beginning of the present state of things; he
admits the possibility of an end.

I have said that the three schools of geological
speculation which I have termed Catastrophism,
Uniformitarianism, and Evolutionism, are commonly supposed to be antagonistic to one another;

and I presume it will have become obvious that in my belief, the last is destined to swallow up the other two. But it is proper to remark that each of the latter has kept alive the tradition of precious truths.

CATASTROPHISM has insisted upon the existence of a practically unlimited bank of force, on which the theorist might draw; and it has cherished the idea of the development of the earth from a state in which its form, and the forces which it exerted, were very different from those we now know. That such difference of form and power once existed is a necessary part of the doctrine of evolution.

UNIFORMITARIANISM, on the other hand, has with equal justice insisted upon a practically unlimited bank of time, ready to discount any quantity of hypothetical paper. It has kept before our eyes the power of the infinitely little, time being granted, and has compelled us to exhaust known causes, before flying to the unknown.

To my mind there appears to be no sort of necessary theoretical antagonism between Catastrophism and Uniformitarianism. On the contrary, it is very conceivable that catastrophes may be part and parcel of uniformity. Let me illustrate my case by analogy. The working of a clock is a model of uniform action; good time-keeping means uniformity of action. But the striking of the clock is essentially a catastrophe; the hammer might be made to blow up a barrel of gunpowder, or turn on a deluge of water; and, by proper arrangement, the clock, instead of marking the hours, might strike at all sorts of irregular periods, never twice alike, in the intervals, force, or number of its blows. Nevertheless, all these irregular, and apparently lawless, catastrophes would be the result of an absolutely uniformitarian action ; and we might have two schools of clock-theorists, one studying the hammer and the other the pendulum.

Still less is there any necessary antagonism between either of these doctrines and that of Evolution, which embraces all that is sound in both Catastrophism and Uniformitarianism, while it rejects the arbitrary assumptions of the one and the, as arbitrary, limitations of the other. Nor is the value of the doctrine of Evolution to the philosophic thinker diminished by the fact that it applies the same method to the living and the not-living world; and embraces, in one stupendous analogy, the growth of a solar system from molecular chaos, the shaping of the earth from the nebulous cubhood of its youth, through innumerable changes and immeasurable ages, to its present form ; and the development of a living being from the shapeless mass of protoplasm we term a germ.

I do not know whether Evolutionism can claim that amount of currency which would entitle it to be called British popular geology ; but, more or less vaguely, it is assuredly present in the minds of most geologists.

Such being the three phases of geological speculation, we are now in position to inquire which of these it is that Sir William Thomson calls upon us to reform in the passages which I have cited.

It is obviously Uniformitarianism which the distinguished physicist takes to be the representative of geological speculation in general. And thus a first issue is raised, inasmuch as many persons (and those not the least thoughtful among the younger geologists) do not accept strict Uniformitarianism as the final form of geological speculation. We should say, if Hutton and Playfair declare the course of the world to have been always the same, point out the fallacy by all means; but, in so doing, do not imagine that you are proving modern geology to be in opposition to natural philosophy. I do not suppose that, at the present day, any geologist would be found to maintain absolute Uniformitarianism, to deny that the rapidity of the rotation of the earth *may* be diminishing, that the sun *may* be waxing dim, or that the earth itself *may* be cooling. Most of us, I suspect, are Gallios, "who care for none of these things," being of opinion that, true or fictitious, they have made no practical difference to the earth, during the period of which a record is preserved in stratified deposits.

The accusation that we have been running counter to the *principles* of natural philosophy, therefore, is devoid of foundation. The only question which can arise is whether we have, or have not, been tacitly making assumptions which are in opposition to certain conclusions which may be drawn from those principles. And this question subdivides itself into two :—the first, are we really contravening such conclusions ? the second, if we are, are those conclusions so firmly based that we may not contravene them ? I reply in the negative to both these questions, and I will give you my reasons for so doing. Sir William Thomson believes that he is able to prove, by physical reasonings, " that the existing state of things on the earth, life on the earth—all geological history showing continuity of life —must be limited within some such period of time as one hundred million years " (*loc. cit.* p. 25).

The first inquiry which arises plainly is, has it ever been denied that this period *may* be enough for the purposes of geology ?

The discussion of this question is greatly embarrassed by the vagueness with which the assumed limit is, I will not say defined, but indicated,—" some such period of past time as one hundred million years." Now does this mean

that it may have been two, or three, or four
hundred million years? Because this really
makes all the difference.[1]

I presume that 100,000 feet may be taken as a
full allowance for the total thickness of stratified
rocks containing traces of life; 100,000 divided
by $100,000,000 = 0\cdot001$. Consequently, the deposit
of 100,000 feet of stratified rock in 100,000,000
years means that the deposit has taken place
at the rate of $\frac{1}{1000}$ of a foot, or, say, $\frac{1}{83}$ of an
inch, per annum.

Well, I do not know that any one is prepared
to maintain that, even making all needful allow-
ances, the stratified rocks may not have been
formed, on the average, at the rate of $\frac{1}{83}$ of an
inch per annum. I suppose that if such could be
shown to be the limit of world-growth, we could
put up with the allowance without feeling that
our speculations had undergone any revolution.
And perhaps, after all, the qualifying phrase
" some such period " may not necessitate the
assumption of more than $\frac{1}{166}$ or $\frac{1}{249}$ or $\frac{1}{332}$ of
an inch of deposit per year, which, of course,
would give us still more ease and comfort.

But, it may be said, that it is biology, and not
geology, which asks for so much time—that the
succession of life demands vast intervals; but

---

[1] Sir William Thomson implies (*loc. cit.* p. 16) that the precise time is of no consequence: "the principle is the same";
but, as the principle is admitted, the whole discussion turns on
its practical results.

this appears to me to be reasoning in a circle. Biology takes her time from geology. The only reason we have for believing in the slow rate of the change in living forms is the fact that they persist through a series of deposits which, geology informs us, have taken a long while to make. If the geological clock is wrong, all the naturalist will have to do is to modify his notions of the rapidity of change accordingly. And I venture to point out that, when we are told that the limitation of the period during which living beings have inhabited this planet to one, two, or three hundred million years requires a complete revolution in geological speculation, the *onus probandi* rests on the maker of the assertion, who brings forward not a shadow of evidence in its support.

Thus, if we accept the limitation of time placed before us by Sir W Thomson, it is not obvious, on the face of the matter, that we shall have to alter, or reform, our ways in any appreciable degree ; and we may therefore proceed with much calmness, and indeed much indifference, as to the result, to inquire whether that limitation is justified by the arguments employed in its support.

These arguments are three in number :—

I. The first is based upon the undoubted fact that the tides tend to retard the rate of the earth's rotation upon its axis. That this must be so is obvious, if one considers, roughly, that the tides result from the pull which the sun and the moon exert upon the sea, causing it to act as a sort of break upon the rotating solid earth.

Kant, who was by no means a mere "abstract philosopher," but a good mathematician and well versed in the physical science of his time, not only proved this in an essay of exquisite clearness and intelligibility, now more than a century old,[1] but deduced from it some of its more important consequences, such as the constant turning of one face of the moon towards the earth.

But there is a long step from the demonstration of a tendency to the estimation of the practical value of that tendency, which is all with which we are at present concerned. The facts bearing on this point appear to stand as follows :—

It is a matter of observation that the moon's mean motion is (and has for the last 3,000 years been) undergoing an acceleration, relatively to the rotation of the earth. Of course this may result from one of two causes: the moon may really have been moving more swiftly in its orbit; or the earth may have been rotating more slowly on its axis.

---

[1] "Untersuchung der Frage ob die Erde in ihrer Umdrehung um die Achse, wodurch sie die Abwechselung des Tages und der Nacht hervorbringt, einige Veränderung seit den ersten Zeiten ihres Ursprunges erlitten habe, &c."—KANT'S *Sämmtliche Werke*, Bd. i. p. 178.

Laplace believed he had accounted for this phenomenon by the fact that the eccentricity of the
earth's orbit has been diminishing throughout
these 3,000 years. This would produce a diminution of the mean attraction of the sun on the
moon; or, in other words, an increase in the
attraction of the earth on the moon; and, consequently, an increase in the rapidity of the orbital
motion of the latter body. Laplace, therefore,
laid the responsibility of the acceleration upon
the moon, and if his views were correct, the tidal
retardation must either be insignificant in amount,
or be counteracted by some other agency.

Our great astronomer, Adams, however, appears
to have found a flaw in Laplace's calculation, and
to have shown that only half the observed retardation could be accounted for in the way he
had suggested. There remains, therefore, the
other half to be accounted for; and here, in the
absence of all positive knowledge, three sets of
hypotheses have been suggested.

(*a.*) M. Delaunay suggests that the earth is at
fault, in consequence of the tidal retardation.
Messrs. Adams, Thomson, and Tait work out this
suggestion, and, "on a certain assumption as to
the proportion of retardations due to the sun and
moon," find the earth may lose twenty-two seconds
of time in a century from this cause.[1]

(*b.*) But M. Dufour suggests that the retardation

---

[1] Sir W. Thomson, *loc. cit.* p. 14.

of the earth (which is hypothetically assumed to exist) may be due in part, or wholly, to the increase of the moment of inertia of the earth by meteors falling upon its surface. This suggestion also meets with the entire approval of Sir W. Thomson, who shows that meteor-dust, accumulating at the rate of one foot in 4,000 years, would account for the remainder of retardation.[1]

(c.) Thirdly, Sir W. Thomson brings forward an hypothesis of his own with respect to the cause of the hypothetical retardation of the earth's rotation :—

"Let us suppose ice to melt from the polar regions (20° round each pole, we may say) to the extent of something more than a foot thick, enough to give 1·1 foot of water over those areas, or 0·006 of a foot of water if spread over the whole globe, which would, in reality, raise the sea-level by only some such undiscoverable difference as three-fourths of an inch or an inch. This, or the reverse, which we believe might happen any year, and could certainly not be detected without far more accurate observations and calculations for the mean sea-level than any hitherto made, would slacken or quicken the earth's rate as a timekeeper by one-tenth of a second per year."[2]

I do not presume to throw the slightest doubt upon the accuracy of any of the calculations made by such distinguished mathematicians as those

---

[1] Sir W. Thomson, *loc. cit.* p. 27.   *Ibid.*

who have made the suggestions I have cited. On the contrary, it is necessary to my argument to assume that they are all correct. But I desire to point out that this seems to be one of the many cases in which the admitted accuracy of mathematical process is allowed to throw a wholly inadmissible appearance of authority over the results obtained by them. Mathematics may be compared to a mill of exquisite workmanship, which grinds you stuff of any degree of fineness; but, nevertheless, what you get out depends upon what you put in; and as the grandest mill in the world will not extract wheat-flour from peascods, so pages of formulæ will not get a definite result out of loose data.

In the present instance it appears to be admitted :—

1. That it is not absolutely certain, after all, whether the moon's mean motion is undergoing acceleration, or the earth's rotation retardation.[1] And yet this is the key of the whole position.

2. If the rapidity of the earth's rotation is diminishing, it is not certain how much of that retardation is due to tidal friction, how much to meteors,—how much to possible excess of melting over accumulation of polar ice, during the period covered by observation, which amounts, at the outside, to not more than 2,600 years.

---

[1] It will be understood that I do not wish to deny that the earth's rotation *may be* undergoing retardation.

3. The effect of a different distribution of land and water in modifying the retardation caused by tidal friction, and of reducing it, under some circumstances, to a minimum, does not appear to be taken into account.

4. During the Miocene epoch the polar ice was certainly many feet thinner than it has been during, or since, the Glacial epoch. Sir W. Thomson tells us that the accumulation of something more than a foot of ice around the poles (which implies the withdrawal of, say, an inch of water from the general surface of the sea) will cause the earth to rotate quicker by one-tenth of a second per annum. It would appear, therefore, that the earth may have been rotating, throughout the whole period which has elapsed from the commencement of the Glacial epoch down to the present time, one, or more, seconds per annum quicker than it rotated during the Miocene epoch.

But, according to Sir W. Thomson's calculation, tidal retardation will only account for a retardation of $22''$ in a century, or $\frac{22}{100}$ (say $\frac{1}{5}$) of a second per annum.

Thus, assuming that the accumulation of polar ice since the Miocene epoch has only been sufficient to produce ten times the effect of a coat of ice one foot thick, we shall have an accelerating cause which covers all the loss from tidal action, and leaves a balance of $\frac{4}{5}$ of a second per annum in the way of acceleration.

If tidal retardation can be thus checked and overthrown by other temporary conditions, what becomes of the confident assertion, based upon the assumed uniformity of tidal retardation, that ten thousand million years ago the earth must have been rotating more than twice as fast as at present, and, therefore, that we geologists are " in direct opposition to the principles of Natural Philosophy " if we spread geological history over that time ?

II. The second argument is thus stated by Sir W. Thomson :—" An article, by myself, published in ' Macmillan's Magazine ' for March 1862, on the age of the sun's heat, explains results of investigation into various questions as to possibilities regarding the amount of heat that the sun could have, dealing with it as you would with a stone, or a piece of matter, only taking into account the sun's dimensions, which showed it to be possible that the sun may have already illuminated the earth for as many as one hundred million years, but at the same time rendered it almost certain that he had not illuminated the earth for five hundred millions of years.   The estimates here are necessarily very vague ; but yet, vague as they are, I do not know that it is possible, upon any reasonable estimate founded on known properties of matter, to say that we can believe the sun has really illuminated the earth for five hundred million years." [1]

[1] *Loc. cit.* p. 20.

I do not wish to "Hansardise" Sir William Thomson by laying much stress on the fact that, only fifteen years ago he entertained a totally different view of the origin of the sun's heat, and believed that the energy radiated from year to year was supplied from year to year—a doctrine which would have suited Hutton perfectly. But the fact that so eminent a physical philosopher has, thus recently, held views opposite to those which he now entertains, and that he confesses his own estimates to be "very vague," justly entitles us to disregard those estimates, if any distinct facts on our side go against them. However, I am not aware that such facts exist. As I have already said, for anything I know, one, two, or three hundred millions of years may serve the needs of geologists perfectly well.

III. The third line of argument is based upon the temperature of the interior of the earth. Sir W Thomson refers to certain investigations which prove that the present thermal condition of the interior of the earth implies either a heating of the earth within the last 20,000 years of as much as 100° F., or a greater heating all over the surface at some time further back than 20,000 years, and then proceeds thus :—

"Now, are geologists prepared to admit that, at some time within the last 20,000 years, there has been all over the earth so high a temperature as that ? I presume not ; no geologist—no *modern*

geologist—would for a moment admit the hypothesis that the present state of underground heat
is due to a heating of the surface at so late a
period as 20,000 years ago. If that is not admitted
we are driven to a greater heat at some time more
than 20,000 years ago. A greater heating all
over the surface than 100° Fahrenheit would kill
nearly all existing plants and animals, I may
safely say. Are modern geologists prepared to
say that all life was killed off the earth 50,000,
100,000, or 200,000 years ago? For the uniformity
theory, the further back the time of high surface-
temperature is put the better; but the further
back the time of heating, the hotter it must have
been. The best for those who draw most largely
on time is that which puts it furthest back; and
that is the theory that the heating was enough to
melt the whole. But even if it was enough to
melt the whole, we must still admit some limit,
such as fifty million years, one hundred million
years, or two or three hundred million years ago.
Beyond that we cannot go." [1]

It will be observed that the "limit" is once
again of the vaguest, ranging from 50,000,000
years to 300,000,000. And the reply is, once
more, that, for anything that can be proved to the
contrary, one or two hundred million years might
serve the purpose, even of a thoroughgoing Hut-
tonian uniformitarian, very well.

[1] *Loc. cit.* p. 24.

But if, on the other hand, the 100,000,000 or 200,000,000 years appear to be insufficient for geological purposes, we must closely criticise the method by which the limit is reached. The argument is simple enough. *Assuming* the earth to be nothing but a cooling mass, the quantity of heat lost per year, *supposing* the rate of cooling to have been uniform, multiplied by any given number of years, will be given the minimum temperature that number of years ago.

But is the earth nothing but a cooling mass, "like a hot-water jar such as is used in carriages," or "a globe of sandstone," and has its cooling been uniform? An affirmative answer to both these questions seems to be necessary to the validity of the calculations on which Sir W. Thomson lays so much stress.

Nevertheless it surely may be urged that such affirmative answers are purely hypothetical, and that other suppositions have an equal right to consideration.

For example, is it not possible that, at the prodigious temperature which would seem to exist at 100 miles below the surface, all the metallic bases may behave as mercury does at a red heat, when it refuses to combine with oxygen; while, nearer the surface, and therefore at a lower temperature, they may enter into combination (as mercury does with oxygen a few degrees below its boiling-point), and so give rise to a heat totally

distinct from that which they possess as cooling
bodies ?   And has it not also been proved by
recent researches that the quality of the atmosphere may immensely affect its permeability to
heat ; and, consequently, profoundly modify the
rate of cooling the globe as a whole ?

I do not think it can be denied that such conditions may exist, and may so greatly affect the
supply, and the loss, of terrestrial heat as to
destroy the value of any calculations which leave
them out of sight.

My functions as your advocate are at an end.   I
speak with more than the sincerity of a mere
advocate when I express the belief that the case
against us has entirely broken down.   The cry for
reform which has been raised without, is superfluous, inasmuch as we have long been reforming
from within, with all needful speed.   And the
critical examination of the grounds upon which
the very grave charge of opposition to the principles
of Natural Philosophy has been brought against
us, rather shows that we have exercised a wise
discrimination in declining, for the present, to
meddle with our foundations.

XI

PALÆONTOLOGY AND THE DOCTRINE
OF EVOLUTION

[1870]

IT is now eight years since, in the absence of
the late Mr. Leonard Horner, who then presided
over us, it fell to my lot, as one of the Secretaries
of this Society, to draw up the customary Annual
Address. I availed myself of the opportunity to
endeavour to " take stock " of that portion of the
science of biology which is commonly called
" palæontology," as it then existed ; and, discussing one after another the doctrines held by
palæontologists, I put before you the results of
my attempts to sift the well-established from
the hypothetical or the doubtful. Permit me
briefly to recall to your minds what those results
were :—

1. The living population of all parts of the
earth's surface which have yet been examined

has undergone a succession of changes which, upon the whole, have been of a slow and gradual character.

2. When the fossil remains which are the evidences of these successive changes, as they have occurred in any two more or less distant parts of the surface of the earth, are compared, they exhibit a certain broad and general parallelism. In other words, certain forms of life in one locality occur in the same general order of succession as, or are *homotaxial* with, similar forms in the other locality.

3. Homotaxis is not to be held identical with synchronism without independent evidence. It is possible that similar, or even identical, faunæ and floræ in two different localities may be of extremely different ages, if the term "age" is used in its proper chronological sense. I stated that "geographical provinces, or zones, may have been as distinctly marked in the Palæozoic epoch as at present; and those seemingly sudden appearances of new genera and species which we ascribe to new creation, may be simple results of migration."

4. The opinion that the oldest known fossils are the earliest forms of life has no solid foundation.

5. If we confine ourselves to positively ascertained facts, the total amount of change in the forms of animal and vegetable life, since the

existence of such forms is recorded, is small.
When compared with the lapse of time since
the first appearance of these forms, the amount
of change is wonderfully small.   Moreover, in
each great group of the animal and vegetable
kingdoms, there are certain forms which I termed
PERSISTENT TYPES, which have remained, with
but very little apparent change, from their first
appearance to the present time.

6. In answer to the question " What, then, does
an impartial survey of the positively ascertained
truths of palæontology testify in relation to the
common doctrines of progressive modification,
which   suppose   that   modification   to   have
taken place by a necessary progress from more
to less embryonic forms, from more to less generalised types, within the limits of the period
represented by the fossiliferous rocks?"   I reply,
' It negatives these doctrines;   for it either
show us no evidence of such modification, or
demonstrates such modification as has occurred
to have been very slight; and, as to the nature
of that modification, it yields no evidence whatsoever that the earlier members of any long-continued group were more generalised in structure
than the later ones."

I think that I cannot employ my last opportunity of addressing you, officially, more properly—
I may say more dutifully—than in revising these
old judgments with such help as further know-

ledge and reflection, and an extreme desire to get
at the truth, may afford me.

1. With respect to the first proposition, I may
remark that whatever may be the case among the
physical geologists, catastrophic palæontologists
are practically extinct. It is now no part of
recognised geological doctrine that the species of
one formation all died out and were replaced by a
brand-new set in the next formation. On the
contrary, it is generally, if not universally, agreed
that the succession of life has been the result of a
slow and gradual replacement of species by species ;
and that all appearances of abruptness of change
are due to breaks in the series of deposits, or other
changes in physical conditions. The continuity of
living forms has been unbroken from the earliest
times to the present day.

2, 3. The use of the word " homotaxis " instead
of " synchronism " has not, so far as I know, found
much favour in the eyes of geologists. I hope,
therefore, that it is a love for scientific caution,
and not mere personal affection for a bantling of
my own, which leads me still to think that the
change of phrase is of importance, and that
the sooner it is made, the sooner shall we get rid
of a number of pitfalls which beset the reasoner
upon the facts and theories of geology.

One of the latest pieces of foreign intelligence
which has reached us is the information that the
Austrian geologists have, at last, succumbed to

the weighty evidence which M. Barrande has accumulated, and have admitted the doctrine of colonies. But the admission of the doctrine of colonies implies the further admission that even identity of organic remains is no proof of the synchronism of the deposits which contain them.

4. The discussions touching the *Eozoon*, which commenced in 1864, have abundantly justified the fourth proposition. In 1862, the oldest record of life was in the Cambrian rocks; but if the *Eozoon* be, as Principal Dawson and Dr Carpenter have shown so much reason for believing, the remains of a living being, the discovery of its true nature carried life back to a period which, as Sir William Logan has observed, is as remote from that during which the Cambrian rocks were deposited, as the Cambrian epoch itself is from the tertiaries. In other words, the ascertained duration of life upon the globe was nearly doubled at a stroke.

5. The significance of persistent types, and of the small amount of change which has taken place even in those forms which can be shown to have been modified, becomes greater and greater in my eyes, the longer I occupy myself with the biology of the past.

Consider how long a time has elapsed since the Miocene epoch. Yet, at that time there is reason to believe that every important group in every order of the *Mammalia* was represented. Even the

comparatively scanty Eocene fauna yields examples
of the orders *Cheiroptera, Insectivora, Rodentia,* and
Perissodactyla ; of *Artiodactyla* under both the
Ruminant and the Porcine modifications ; of *Carni-*
vora, Cetacea, and *Marsupialia.*

Or, if we go back to the older half of the Mesozoic epoch, how truly surprising it is to find
every order of the *Reptilia,* except the *Ophidia,*
represented ; while some groups, such as the
Ornithoscelida and the *Pterosauria,* more specialised
than any which now exist, abounded.

There is one division of the *Amphibia* which
offers especially important evidence upon this
point, inasmuch as it bridges over the gap between
the Mesozoic and the Palæozoic formations (often
supposed to be of such prodigious magnitude), extending, as it does, from the bottom of the Carboniferous series to the top of the Trias, if not
into the Lias. I refer to the Labyrinthodonts.
As the Address of 1862 was passing through the
press, I was able to mention, in a note, the
discovery of a large Labyrinthodont, with well-
ossified vertebræ, in the Edinburgh coal-field.
Since that time eight or ten distinct genera of
Labyrinthodonts have been discovered in the
Carboniferous rocks of England, Scotland, and
Ireland, not to mention the American forms
described by Principal Dawson and Professor
Cope. So that, at the present time, the Labyrin-
thodont Fauna of the Carboniferous rocks is more

extensive and diversified than that of the Trias, while its chief types, so far as osteology enables us to judge, are quite as highly organised. Thus it is certain that a comparatively highly organised vertebrate type, such as that of the Labyrintho- donts, is capable of persisting, with no considerable change, through the period represented by the vast deposits which constitute the Carboniferous, the Permian, and the Triassic formations.

The very remarkable results which have been brought to light by the sounding and dredging operations, which have been carried on with such remarkable success by the expeditions sent out by our own, the American, and the Swedish Govern- ments, under the supervision of able naturalists, have a bearing in the same direction. These in- vestigations have demonstrated the existence, at great depths in the ocean, of living animals in some cases identical with, in others very similar to, those which are found fossilised in the white chalk. The *Globigerinæ*, Cyatholiths, Cocco- spheres, Discoliths in the one are absolutely identical with those in the other ; there are identical, or closely analogous, species of Sponges, Echinoderms, and Brachiopods. Off the coast of Portugal, there now lives a species of *Beryx*, which, doubtless, leaves its bones and scales here and there in the Atlantic ooze, as its predecessor left its spoils in the mud of the sea of the Cretaceous epoch.

Many years ago[1] I ventured to speak of the Atlantic mud as "modern chalk," and I know of no fact inconsistent with the view which Professor Wyville Thomson has advocated, that the modern chalk is not only the lineal descendant of the ancient chalk, but that it remains, so to speak, in the possession of the ancestral estate; and that from the Cretaceous period (if not much earlier) to the present day, the deep sea has covered a large part of what is now the area of the Atlantic. But if *Globigerina*, and *Terebratula caput-serpentis* and *Beryx*, not to mention other forms of animals and of plants, thus bridge over the interval between the present and the Mesozoic periods, is it possible that the majority of other living things underwent a "sea-change into something new and strange" all at once?

6. Thus far I have endeavoured to expand and to enforce by fresh arguments, but not to modify in any important respect, the ideas submitted to you on a former occasion. But when I come to the propositions touching progressive modification, it appears to me, with the help of the new light which has broken from various quarters, that there is much ground for softening the somewhat Brutus-like severity with which, in 1862, I dealt with a doctrine, for the truth of which I should have been glad enough to be able to find a good

---

[1] See an article in the *Saturday Review*, for 1858, on "Chalk, Ancient and Modern."

foundation. So far, indeed, as the *Invertebrata* and the lower *Vertebrata* are concerned, the facts and the conclusions which are to be drawn from them appear to me to remain what they were. For anything that, as yet, appears to the contrary, the earliest known Marsupials may have been as highly organised as their living congeners; the Permian lizards show no signs of inferiority to those of the present day; the Labyrinthodonts cannot be placed below the living Salamander and Triton; the Devonian Ganoids are closely related to *Polypterus* and to *Lepidosiren*.

But when we turn to the higher *Vertebrata*, the results of recent investigations, however we may sift and criticise them, seem to me to leave a clear balance in favour of the doctrine of the evolution of living forms one from another. Nevertheless, in discussing this question, it is very necessary to discriminate carefully between the different kinds of evidence from fossil remains which are brought forward in favour of evolution.

Every fossil which takes an intermediate place between forms of life already known, may be said, so far as it is intermediate, to be evidence in favour of evolution, inasmuch as it shows a possible road by which evolution may have taken place. But the mere discovery of such a form does not, in itself, prove that evolution took place by and through it, nor does it constitute more than presumptive evidence in favour of evolution in general. Suppose A, B, C to be three forms, while B is intermediate in structure between A and C. Then the doctrine of evolution offers four possible alternatives. A may have become C by way of B; or C may have become A by way of B; or A and C may be independent modifications of B; or A, B, and C may be independent modifications of some unknown D Take the case of the Pigs, the *Anoplotheridæ*, and the Ruminants. The *Anoplotheridæ* are intermediate between the first and the last; but this does not tell us whether the Ruminants have come from the Pigs, or the Pigs from Ruminants, or both from *Anoplotheridæ*, or whether Pigs, Ruminants, and *Anoplotheridæ* alike may not have diverged from some common stock.

But if it can be shown that A, B, and C exhibit successive stages in the degree of modification, or specialisation, of the same type; and if, further, it can be proved that they occur in successively newer deposits, A being in the oldest and C in the newest, then the intermediate character of B has quite another importance, and I should accept it, without hesitation, as a link in the genealogy of C. I should consider the burden of proof to be thrown upon any one who denied C to have been derived from A by way of B, or in some closely analogous fashion; for it is always probable that one may not hit upon the exact line of filiation, and, in dealing with fossils, may mistake uncles and nephews for fathers and sons.

I think it necessary to distinguish between the former and the latter classes of intermediate forms, as *intercalary types* and *linear types*. When I apply the former term, I merely mean to say that as a matter of fact, the form B, so named, is intermediate between the others, in the sense in which the *Anoplotherium* is intermediate between the Pigs and the Ruminants—without either affirming, or denying, any direct genetic relation between the three forms involved. When I apply the latter term, on the other hand, I mean to express the opinion that the forms A, B, and C constitute a line of descent, and that B is thus part of the lineage of C.

From the time when Cuvier's wonderful researches upon the extinct Mammals of the Paris gypsum first made intercalary types known, and caused them to be recognised as such, the number of such forms has steadily increased among the higher *Mammalia*. Not only do we now know numerous intercalary forms of *Ungulata*, but M. Gaudry's great monograph upon the fossils of Pikermi (which strikes me as one of the most˙ perfect pieces of palæontological work I have seen for a long time) shows us, among the *Primates*, *Mesopithecus* as an intercalary form between the *Semnopitheci* and the *Macaci;* and among the *Carnivora, Hyœnictis* and *Ictitherium* as intercalary, or, perhaps, linear types between the *Viverridæ* and the *Hyænidæ*.

Hardly any order of the higher *Mammalia* stands so apparently separate and isolated from the rest as that of the *Cetacea* ; though a careful consideration of the structure of the pinnipede *Carnivora*, or Seals, shows, in them, many an approximation towards the still more completely marine mammals. The extinct *Zeuglodon*, however, presents us with an intercalary form between the type of the Seals and that of the Whales. The skull of this great Eocene sea-monster, in fact, shows by the narrow and prolonged inter-orbital region ; the extensive union of the parietal bones in a sagittal suture ; the well-developed nasal bones ; the distinct and large incisors implanted in premaxillary bones, which take a full share in bounding the fore part of the gape ; the two-fanged molar teeth with triangular and serrated crowns, not exceeding five on each side in each jaw ; and the existence of a deciduous dentition—its close relation with the Seals. While, on the other hand, the produced rostral form of the snout, the long symphysis, and the low coronary process of the mandible are approximations to the cetacean form of those parts.

The scapula resembles that of the cetacean *Hyperoodon*, but the supra-spinous fossa is larger and more seal-like ; as is the humerus, which differs from that of the *Cetacea* in presenting true

articular surfaces for the free jointing of the
bones of the fore-arm. In the apparently complete absence of hinder limbs, and in the characters
of the vertebral column, the *Zeuglodon* lies on the
cetacean side of the boundary line ; so that upon
the whole, the Zeuglodonts, transitional as they
are, are conveniently retained in the cetacean
order. And the publication, in 1864, of M. Van
Beneden's memoir on the Miocene and Pliocene
Squalodon, furnished much better means than
anatomists previously possessed of fitting in
another link of the chain which connects the
existing *Cetacea* with *Zeuglodon*. The teeth are
much more numerous, although the molars exhibit
the zeuglodont double fang ; the nasal bones are
very short, and the upper surface of the rostrum
presents the groove, filled up during life by the
prolongation of the ethmoidal cartilage, which is
so characteristic of the majority of the *Cetacea*.

It appears to me that, just as among the
existing *Carnivora*, the walruses and the eared
seals are intercalary forms between the fissipede
Carnivora and the ordinary seals, so the Zeuglodonts are intercalary between the *Carnivora*, as a
whole, and the *Cetacea*. Whether the Zeuglodonts
are also linear types in their relation to these two
groups cannot be ascertained, until we have more
definite knowledge than we possess at present,
respecting the relations in time of the *Carnivora*
and *Cetacea*.

Thus far we have been concerned with the intercalary types which occupy the intervals between Families or Orders of the same class; but the investigations which have been carried on by Professor Gegenbaur, Professor Cope, and myself into the structure and relations of the extinct reptilian forms of the *Ornithoscelida* (or *Dinosauria* and *Compsognatha*) have brought to light the existence of intercalary forms between what have hitherto been always regarded as very distinct classes of the vertebrate sub-kingdom, namely *Reptilia* and *Aves*. Whatever inferences may, or may not, be drawn from the fact, it is now an established truth that, in many of these *Ornithoscelida*, the hind limbs and the pelvis are much more similar to those of Birds than they are to those of Reptiles, and that these Bird-reptiles, or Reptile-birds, were more or less completely bipedal.

When I addressed you in 1862, I should have been bold indeed had I suggested that palæontology would before long show us the possibility of a direct transition from the type of the lizard to that of the ostrich. At the present moment, we have, in the *Ornithoscelida*, the intercalary type, which proves that transition to be something more than a possibility; but it is very doubtful whether any of the genera of *Ornithoscelida* with which we are at present acquainted are the actual linear types by which the transition from the

lizard to the bird was effected. These, very probably, are still hidden from us in the older formations.

Let us now endeavour to find some cases of
true linear types, or forms which are intermediate
between others because they stand in a direct
genetic relation to them. It is no easy matter to
find clear and unmistakable evidence of filiation
among fossil animals; for, in order that such
evidence should be quite satisfactory, it is necessary
that we should be acquainted with all the most
important features of the organisation of the
animals which are supposed to be thus related, and
not merely with the fragments upon which the
genera and species of the palæontologist are so
often based. M. Gaudry has arranged the species
of *Hyænidæ, Proboscidea, Rhinocerotidæ,* and *Equidæ*
in their order of filiation from their earliest appearance in the Miocene epoch to the present time, and
Professor Rutimeyer has drawn up similar schemes
for the Oxen and other *Ungulata*—with what, I
am disposed to think, is a fair and probable approximation to the order of nature. But, as no one is
better aware than these two learned, acute, and
philosophical biologists, all such arrangements
must be regarded as provisional, except in those
cases in which, by a fortunate accident, large
series of remains are obtainable from a thick and
widespread series of deposits. It is easy to
accumulate probabilities—hard to make out some

particular case in such a way that it will stand rigorous criticism.

After much search, however, I think that such a case is to be made out in favour of the pedigree of the Horses.

The genus *Equus* is represented as far back as the latter part of the Miocene epoch; but in deposits belonging to the middle of that epoch its place is taken by two other genera, *Hipparion* and *Anchitherium;* [1] and, in the lowest Miocene and upper Eocene, only the last genus occurs. A species of *Anchitherium* was referred by Cuvier to the *Palæotheria* under the name of *P. aurelianense.* The grinding-teeth are in fact very similar in shape and in pattern, and in the absence of any thick layer of cement, to those of some species of *Palæotherium*, especially Cuvier's *Palæotherium minus*, which has been formed into a separate genus, *Plagiolophus*, by Pomel. But in the fact that there are only six full-sized grinders in the lower jaw, the first premolar being very small; that the anterior grinders are as large as, or rather larger than, the posterior ones; that the

---

[1] Hermann von Meyer gave the name of *Anchitherium* to *A. Ezquerræ;* and in his paper on the subject he takes great pains to distinguish the latter as the type of a new genus, from Cuvier's *Palæotherium d'Orléans.* But it is precisely the *Palæotherium d'Orléans* which is the type of Christol's genus *Hipparitherium;* and thus, though *Hipparitherium* is of later date than *Anchitherium*, it seemed to me to have a sort of equitable right to recognition when this Address was written. On the whole, however, it seems most convenient to adopt *Anchitherium.*

second premolar has an anterior prolongation ; and that the posterior molar of the lower jaw has, as Cuvier pointed out, a posterior lobe of much smaller size and different form, the dentition of *Anchitherium* departs from the type of the *Palæotherium*, and approaches that of the Horse.

Again, the skeleton of *Anchitherium* is extremely equine. M. Christol goes so far as to say that the description of the bones of the horse, or the ass, current in veterinary works, would fit those of *Anchitherium*. And, in a general way, this may be true enough ; but there are some most important differences, which, indeed, are justly indicated by the same careful observer. Thus the ulna is complete throughout, and its shaft is not a mere rudiment, fused into one bone with the radius. There are three toes, one large in the middle and one small on each side. The femur is quite like that of a horse, and has the characteristic fossa above the external condyle. In the British Museum there is a most instructive specimen of the leg-bones, showing that the fibula was represented by the external malleolus and by a flat tongue of bone, which extends up from it on the outer side of the tibia, and is closely ankylosed with the latter bone.[1] The hind toes

[1] I am indebted to M. Gervais for a specimen which indicates that the fibula was complete, at any rate, in some cases ; and for a very interesting ramus of a mandible, which shows that, as in the *Palæotheria*, the hindermost milk-molar of the lower

are three, like those of the fore leg; and the
middle metatarsal bone is much less compressed
from side to side than that of the horse.

In the *Hipparion*, the teeth nearly resemble
those of the Horses, though the crowns of the
grinders are not so long; like those of the Horses,
they are abundantly coated with cement. The
shaft of the ulna is reduced to a mere style, ankylosed throughout nearly its whole length with the
radius, and appearing to be little more than a
ridge on the surface of the latter bone until it is
carefully examined. The front toes are still three,
but the outer ones are more slender than in
Anchitherium, and their hoofs smaller in proportion
to that of the middle toe; they are, in fact, reduced to mere dew-claws, and do not touch the
ground. In the leg, the distal end of the fibula is
so completely united with the tibia that it appears
to be a mere process of the latter bone, as in the
Horses.

In *Equus*, finally, the crowns of the grinding-
teeth become longer, and their patterns are slightly
modified; the middle of the shaft of the ulna
usually vanishes, and its proximal and distal ends
ankylose with the radius. The phalanges of the
two outer toes in each foot disappear, their metacarpal and metatarsal bones being left as the
" splints."

jaw was devoid of the posterior lobe which exists in the hindermost true molar.

The *Hipparion* has large depressions on the face in front of the orbits, like those for the " larmiers " of many ruminants ; but traces of these are to be seen in some of the fossil horses from the Sewalik Hills; and, as Leidy's recent researches show, they are preserved in *Anchitherium.*

When we consider these facts, and the further circumstance that the Hipparions, the remains of which have been collected in immense numbers, were subject, as M. Gaudry and others have pointed out, to a great range of variation, it appears to me impossible to resist the conclusion that the types of the *Anchitherium,* of the *Hipparion,* and of the ancient Horses constitute the lineage of the modern Horses, the *Hipparion* being the intermediate stage between the other two, and answering to B in my former illustration.

The process by which the *Anchitherium* has been converted into *Equus* is one of specialisation, or of more and more complete deviation from what might be called the average form of an ungulate mammal. In the Horses, the reduction of some parts of the limbs, together with the special modification of those which are left, is carried to a greater extent than in any other hoofed mammals. The reduction is less and the specialisation is less in the *Hipparion,* and still less in the *Anchitherium ;* but yet, as compared with other mammals, the reduction and specialisation of parts in the *Anchitherium* remain great.

Is it not probable then, that, just as in the Miocene epoch, we find an ancestral equine form less modified than *Equus*, so, if we go back to the Eocene epoch, we shall find some quadruped related to the *Anchitherium*, as *Hipparion* is related to *Equus*, and consequently departing less from the average form?

I think that this desideratum is very nearly, if not quite, supplied by *Plagiolophus*, remains of which occur abundantly in some parts of the Upper and Middle Eocene formations. The patterns of the grinding-teeth of *Plagiolophus* are similar to those of *Anchitherium*, and their crowns are as thinly covered with cement; but the grinders diminish in size forwards, and the last lower molar has a large hind lobe, convex outwards and concave inwards, as in *Palæotherium*. The ulna is complete and much larger than in any of the *Equidæ*, while it is more slender than in most of the true *Palæotheria*; it is fixedly united, but not ankylosed, with the radius. There are three toes in the fore limb, the outer ones being slender, but less attenuated than in the *Equidæ*. The femur is more like that of the *Palæotheria* than that of the horse, and has only a small depression above its outer condyle in the place of the great fossa which is so obvious in the *Equidæ*. The fibula is distinct, but very slender, and its distal end is ankylosed with the tibia. There are three toes on the hind foot having similar proportions to those on the fore foot. The principal metacarpal and metatarsal bones are flatter than they are in any of the *Equidæ*; and the metacarpal bones are longer than the metatarsals, as in the *Palæotheria*.

In its general form, *Plagiolophus* resembles a very small and slender horse,[1] and is totally unlike the reluctant, pig-like creature depicted in Cuvier's restoration of his *Palæotherium minus* in the "Ossemens Fossiles."

It would be hazardous to say that *Plagiolophus* is the exact radical form of the Equine quadrupeds; but I do not think there can be any reasonable doubt that the latter animals have resulted from the modification of some quadruped similar to *Plagiolophus*.

We have thus arrived at the Middle Eocene formation, and yet have traced back the Horses only to a three-toed stock; but these three-toed forms, no less than the Equine quadrupeds themselves, present rudiments of the two other toes which appertain to what I have termed the "average" quadruped. If the expectation raised by the splints of the Horses that, in some ancestor of the Horses, these splints would be found to be complete digits, has been verified, we are fur-

---

[1] Such, at least, is the conclusion suggested by the proportions of the skeleton figured by Cuvier and De Blainville; but perhaps something between a Horse and an Agouti would be nearest the mark.

nished with very strong reasons for looking for a
no less complete verification of the expectation
that the three-toed *Plagiolophus*-like " avus " of the
horse must have had a five-toed " atavus " at some
earlier period.

No such five-toed " atavus," however, has yet
made its appearance among the few middle and
older Eocene *Mammalia* which are known.

Another series of closely affiliated forms, though
the evidence they afford is perhaps less complete
than that of the Equine series, is presented to
us by the *Dichobune* of the Eocene epoch, the
Cainotherium of the Miocene, and the *Tragulidæ*,
or so-called " Musk-deer," of the present day.

The *Tragulidæ* have no incisors in the upper
jaw, and only six grinding-teeth on each side of
each jaw ; while the canine is moved up to the
outer incisor, and there is a diastema in the lower
jaw.   There are four complete toes on the hind
foot, but the middle metatarsals usually become,
sooner or later, ankylosed into a cannon bone.
The navicular and the cuboid unite, and the
distal end of the fibula is ankylosed with the
tibia.

In *Cainotherium* and *Dichobune* the upper
incisors are fully developed. There are seven
grinders ; the teeth form a continuous series without a diastema.   The metatarsals, the navicular
and cuboid, and the distal end of the fibula,
remain free. In the *Cainotherium*, also, the second

metacarpal is developed, but is much shorter than the third, while the fifth is absent or rudimentary. In this respect it resembles *Anoplotherium secundarium*. This circumstance, and the peculiar pattern of the upper molars in *Cainotherium*, lead me to hesitate in considering it as the actual ancestor of the modern *Tragulidæ*. If *Dichobune* has a fore-toed fore foot (though I am inclined to suspect that it resembles *Cainotherium*), it will be a better representative of the oldest forms of the Traguline series ; but *Dichobune* occurs in the Middle Eocene, and is, in fact, the oldest known artiodactyle mammal. Where, then, must we look for its five-toed ancestor ?

If we follow down other lines of recent and tertiary *Ungulata*, the same question presents itself. The Pigs are traceable back through the Miocene epoch to the Upper Eocene, where they appear in the two well-marked forms of *Hyopopotamus* and *Chœropotamus ;* but *Hyopotamus* appears to have had only two toes.

Again, all the great groups of the Ruminants, the *Bovidæ*, *Antilopidæ*, *Camelopardalidæ*, and *Cervidæ*, are represented in the Miocene epoch, and so are the Camels. The Upper Eocene *Anoplotherium*, which is intercalary between the Pigs and the *Tragulidæ*, has only two, or, at most, three toes. Among the scanty mammals of the Lower Eocene formation we have the perissodactyle *Ungulata* represented by *Coryphodon*,

Hyracotherium, and *Pliolophus*. Suppose for a moment, for the sake of following out the argument, that *Pliolophus* represents the primary stock of the Perissodactyles, and *Dichobune* that of the Artiodactyles (though I am far from saying that such is the case), then we find, in the earliest fauna of the Eocene epoch to which our investigations carry us, the two divisions of the *Ungulata* completely differentiated, and no trace of any common stock of both, or of five-toed predecessors to either. With the case of the Horses before us, justifying a belief in the production of new animal forms by modification of old ones, I see no escape from the necessity of seeking for these ancestors of the *Ungulata* beyond the limits of the Tertiary formations.

I could as soon admit special creation, at once, as suppose that the Perissodactyles and Artiodactyles had no five-toed ancestors. And when we consider how large a portion of the Tertiary period elapsed before *Anchitherium* was converted into *Equus*, it is difficult to escape the conclusion that a large proportion of time anterior to the Tertiary period must have been expended in converting the common stock of the *Ungulata* into Perissodactyles and Artiodactyles.

The same moral is inculcated by the study of every other order of Tertiary monodelphous *Mammalia*. Each of these orders is represented in the Miocene epoch: the Eocene formation, as

I have already said, contains *Cheiroptera, Insectivora, Rodentia, Ungulata, Carnivora,* and *Cetacea.*
But the *Cheiroptera* are extreme modifications
of the *Insectivora,* just as the *Cetacea* are extreme
modifications of the Carnivorous type ; and therefore it is to my mind incredible that monodelphous
Insectivora and *Carnivora* should not have been
abundantly developed, along with *Ungulata,* in
the Mesozoic epoch. But if this be the case,
how much further back must we go to find the
common stock of the monodelphous *Mammalia?*
As to the *Didelphia,* if we may trust the evidence
which seems to be afforded by their very scanty
remains, a Hypsiprymnoid form existed at the
epoch of the Trias, contemporaneously with a
Carnivorous form. At the epoch of the Trias,
therefore, the *Marsupialia* must have already
existed long enough to have become differentiated
into carnivorous and herbivorous forms. But the
Monotremata are lower forms than the *Didelphia*
which last are intercalary between the *Ornitho-
delphia* and the *Monodelphia.* To what point of
the Palæozoic epoch, then, must we, upon any
rational estimate, relegate the origin of the
Monotremata?

The investigation of the occurrence of the
classes and of the orders of the *Sauropsida* in time
points in exactly the same direction. If, as there
is great reason to believe, true Birds existed in
the Triassic epoch, the ornithoscelidous forms by

which Reptiles passed into Birds must have preceded them.   In fact there is, even at present,
considerable ground for suspecting the existence
of *Dinosauria* in the Permian formations; but, in
that case, lizards must be of still earlier date.
And if the very small differences which are
observable between the *Crocodilia* of the older
Mesozoic formations and those of the present day
furnish any sort of approximation towards an
estimate of the average rate of change among the
Sauropsida, it is almost appalling to reflect how far
back in Palæozoic times we must go, before we
can hope to arrive at that common stock from
which the *Crocodilia*, *Lacertilia*, *Ornithoscelida*,
and *Plesiosauria*, which had attained so great a
development in the Triassic epoch, must have
been derived.

The *Amphibia* and *Pisces* tell the same story.
There is not a single class of vertebrated animals
which, when it first appears, is represented by
analogues of the lowest known members of the
same class.   Therefore, if there is any truth in
the doctrine of evolution, every class must be vastly
older than the first record of its appearance upon
the surface of the globe.   But if considerations of
this kind compel us to place the origin of vertebrated animals at a period sufficiently distant
from the Upper Silurian, in which the first Elas-
mobranchs and Ganoids occur, to allow of the
evolution of such fishes as these from a Vertebrate

as simple as the *Amphioxus*, I can only repeat
that it is appalling to speculate upon the extent
to which that origin must have preceded the
epoch of the first recorded appearance of vertebrate life.

Such is the further commentary which I have
to offer upon the statement of the chief results of
palæontology which I formerly ventured to lay
before you.

But the growth of knowledge in the interval
makes me conscious of an omission of considerable
moment in that statement, inasmuch as it contains
no reference to the bearings of palæontology upon
the theory of the distribution of life; nor takes
note of the remarkable manner in which the facts
of distribution, in present and past times, accord
with the doctrine of evolution, especially in regard
to land animals.

That connection between palæontology and
geology and the present distribution of terrestrial
animals, which so strikingly impressed Mr. Darwin,
thirty years ago, as to lead him to speak of a " law
of succession of types," and of the wonderful re-
lationship on the same continent between the
dead and the living, has recently received much
elucidation from the researches of Gaudry, of
Rütimeyer, of Leidy, and of Alphonse Milne-
Edwards, taken in connection with the earlier
labours of our lamented colleague Falconer; and

it has been instructively discussed in the thoughtful and ingenious work of Mr. Andrew Murray
" On the Geographical Distribution of Mammals." [1]

I propose to lay before you, as briefly as I can,
the ideas to which a long consideration of the
subject has given rise in my mind.

If the doctrine of evolution is sound, one of its
immediate consequences clearly is, that the present
distribution of life upon the globe is the product
of two factors, the one being the distribution
which obtained in the immediately preceding
epoch, and the other the character and the extent
of the changes which have taken place in physical
geography between the one epoch and the other;
or, to put the matter in another way, the Fauna
and Flora of any given area, in any given epoch,
can consist only of such forms of life as are directly
descended from those which constituted the Fauna
and Flora of the same area in the immediately
preceding epoch, unless the physical geography
(under which I include climatal conditions) of
the area has been so altered as to give rise to
immigration of living forms from some other
area.

The evolutionist, therefore, is bound to grapple

[1] The paper "On the Form and Distribution of the Land-
tracts during the Secondary and Tertiary Periods respectively;
and on the Effect upon Animal Life which great Changes in
Geographical Configuration have probably produced," by Mr.
Searles V. Wood, jun., which was published in the *Philosophical
Magazine*, in 1862, was unknown to me when this Address
was written.   It is well worthy of the most careful study.

with the following problem whenever it is clearly put before him :—Here are the Faunæ of the same area during successive epochs. Show good cause for believing either that these Faunæ have been derived from one another by gradual modification, or that the Faunæ have reached the area in question by migration from some area in which they have undergone their development.

I propose to attempt to deal with this problem, so far as it is exemplified by the distribution of the terrestrial *Vertebrata*, and I shall endeavour to show you that it is capable of solution in a sense entirely favourable to the doctrine of evolution.

I have elsewhere [1] stated at length the reasons which lead me to recognise four primary distributional provinces for the terrestrial *Vertebrata* in the present world, namely,—first, the *Novozelanian*, or New-Zealand province; secondly, the *Australian* province, including Australia, Tasmania, and the Negrito Islands; thirdly, *Austro-Columbia*, or South America *plus* North America as · far as Mexico; and fourthly, the rest of the world, or *Arctogœa*, in which province America north of Mexico constitutes one sub-province, Africa south of the Sahara a second, Hindostan a third, and the remainder of the Old World a fourth.

Now the truth which Mr. Darwin perceived and

---

[1] " On the Classification and Distribution of the Alectoro-morphæ ; " *Proceedings of the Zoological Society*, 1868.

promulgated as "the law of the succession of types" is, that, in all these provinces, the animals found in Pliocene or later deposits are closely affined to those which now inhabit the same provinces; and that, conversely, the forms characteristic of other provinces are absent. North and South America, perhaps, present one or two exceptions to the last rule, but they are readily susceptible of explanation. Thus, in Australia, the later Tertiary mammals are marsupials (possibly with the exception of the Dog and a Rodent or two, as at present). In Austro-Columbia, the later Tertiary fauna exhibits numerous and varied forms of Platyrrhine Apes, Rodents, Cats, Dogs, Stags, *Edentata*, and Opossums; but, as at present, no Catarrhine Apes, no Lemurs, no *Insectivora*, Oxen, Antelopes, Rhinoceroses, nor *Didelphia* other than Opossums. And in the widespread Arctogæal province, the Pliocene and later mammals belong to the same groups as those which now exist in the province. The law of succession of types, therefore, holds good for the present epoch as compared with its predecessor. Does it equally well apply to the Pliocene fauna when we compare it with that of the Miocene epoch? By great good fortune, an extensive mammalian fauna of the latter epoch has now become known, in four very distant portions of the Arctogæal province which do not differ greatly in latitude. Thus Falconer and Cautley have made known the

fauna of the sub-Himalayas and the Perim Islands;
Gaudry that of Attica; many observers that of
Central Europe and France; and Leidy that of
Nebraska, on the eastern flank of the Rocky
Mountains. The results are very striking. The
total Miocene fauna comprises many genera and
species of Catarrhine Apes, of Bats, of *Insectivora*;
of Arctogæal types of *Rodentia*; of *Proboscidea*; of
equine, rhinocerotic, and tapirine quadrupeds; of
cameline, bovine, antilopine, cervine, and traguline
Ruminants; of Pigs and Hippopotamuses; of
Viverridæ and *Hyænidæ* among other *Carnivora*;
with *Edentata* allied to the Arctogæal *Orycteropus*
and *Manis*, and not to the Austro-Columbian
Edentates. The only type present in the Miocene,
but absent in the existing, fauna of Eastern Arctogæa, is that of the *Didelphidæ*, which, however,
remains in North America.

But it is very remarkable that while
Miocene fauna of the Arctogæal province, as
a whole, is of the same character as the existing
fauna of the same province, as a whole, the component elements of the fauna were differently associated. In the Miocene epoch, North America
possessed Elephants, Horses, Rhinoceroses, and a
great number and variety of Ruminants and Pigs,
which are absent in the present indigenous fauna;
Europe had its Apes, Elephants, Rhinoceroses,
Tapirs, Musk-deer, Giraffes, Hyænas, great Cats,
Edentates, and Opossum-like Marsupials, which

XI   PALÆONTOLOGY AND EVOLUTION   371

have equally vanished from its present fauna; and in Northern India, the African types of Hippopotamuses, Giraffes, and Elephants were mixed up with what are now the Asiatic types of the latter, and with Camels, and Semnopithecine and Pithecine Apes of no less distinctly Asiatic forms.

In fact the Miocene mammalian fauna of Europe and the Himalayan regions contains, associated together, the types which are at present separately located in the South-African and Indian sub-provinces of Arctogæa. Now there is every reason to believe, on other grounds, that both Hindostan, south of the Ganges, and Africa, south of the Sahara, were separated by a wide sea from Europe and North Asia during the Middle and Upper Eocene epochs. Hence it becomes highly probable that the well-known similarities, and no less remarkable differences between the present Faunæ of India and South Africa have arisen in some such fashion as the following. Some time during the Miocene epoch, possibly when the Himalayan chain was elevated, the bottom of the nummulitic sea was upheaved and converted into dry land, in the direction of a line extending from Abyssinia to the mouth of the Ganges. By this means, the Dekhan on the one hand, and South Africa on the other, became connected with the Miocene dry land and with one another. The Miocene mammals spread gradually over this intermediate

dry land; and if the condition of its eastern and
western ends offered as wide contrasts as the
valleys of the Ganges and Arabia do now, many
forms which made their way into Africa must
have been different from those which reached
the Dekhan, while others might pass into both
these sub-provinces.

That there was a continuity of dry land between
Europe and North America during the Miocene
epoch, appears to me to be a necessary consequence
of the fact that many genera of terrestrial
mammals, such as *Castor, Hystrix, Elephas,
Mastodon, Equus, Hipparion, Anchitherium, Rhinoceros, Cervus, Amphicyon, Hyænarctos,* and *Machairodus,* are common to the Miocene formations of
the two areas, and have as yet been found (except
perhaps *Anchitherium*) in no deposit of earlier age.
Whether this connection took place by the east,
or by the west, or by both sides of the Old
World, there is at present no certain evidence, and
the question is immaterial to the present argument; but, as there are good grounds for the
belief that the Australian province and the Indian
and South-African sub-provinces were separated
by sea from the rest of Arctogæa before the
Miocene epoch, so it has been rendered no less
probable, by the investigations of Mr. Carrick
Moore and Professor Duncan, that Austro-Columbia
was separated by sea from North America during
a large part of the Miocene epoch.

It is unfortunate that we have no knowledge of the Miocene mammalian fauna of the Australian and Austro-Columbian provinces; but, seeing that not a trace of a Platyrrhine Ape, of a Procyonine Carnivore, of a characteristically South-American Rodent, of a Sloth, an Armadillo, or an Ant-eater has yet been found in Miocene deposits of Arctogæa, I cannot doubt that they already existed in the Miocene Austro-Columbian province.

Nor is it less probable that the characteristic types of Australian Mammalia were already developed in that region in Miocene times.

But Austro-Columbia presents difficulties from which Australia is free; *Camelidæ* and *Tapiridæ* are now indigenous in South America as they are in Arctogæa; and, among the Pliocene Austro-Columbian mammals, the Arctogæal genera *Equus*, *Mastodon*, and *Machairodus* are numbered. Are these Postmiocene immigrants, or Præmiocene natives?

Still more perplexing are the strange and interesting forms *Toxodon*, *Macrauchenia*, *Typotherium*, and a new Anoplotherioid mammal (*Homalodotherium*) which Dr. Cunningham sent over to me some time ago from Patagonia. I confess I am strongly inclined to surmise that these last, at any rate, are remnants of the population of Austro-Columbia before the Miocene epoch, and were not derived from Arctogæa by way of the north and east.

The fact that this immense fauna of Miocene Arctogæa is now fully and richly represented only in India and in South Africa, while it is shrunk and depauperised in North Asia, Europe, and North America, becomes at once intelligible, if we suppose that India and South Africa had but a scanty mammalian population before the Miocene immigration, while the conditions were highly favourable to the new comers.   It is to be supposed that these new regions offered themselves to the Miocene Ungulates, as South America and Australia offered themselves to the cattle, sheep, and horses of modern colonists.   But, after these great areas were thus peopled, came the Glacial epoch, during which the excessive cold, to say nothing of depression and ice-covering, must have almost depopulated all the northern parts of Arctogæa, destroying all the higher mammalian forms, except those which, like the Elephant and Rhinoceros, could adjust their coats to the altered conditions.   Even these must have been. driven away from the greater part of the area; only those Miocene mammals which had passed into Hindostan and into South Africa would escape decimation by such changes in the physical geography of Arctogæa. And when the northern hemisphere passed into its present condition, these lost tribes of the Miocene Fauna were hemmed by the Himalayas, the Sahara, the Red Sea, and the Arabian deserts, within their present boundaries.

Now, on the hypothesis of evolution, there is no sort of difficulty in admitting that the differences between the Miocene forms of the mammalian Fauna and those which exist at present are the results of gradual modification; and, since such differences in distribution as obtain are readily explained by the changes which have taken place in the physical geography of the world since the Miocene epoch, it is clear that the result of the comparison of the Miocene and present Faunæ is distinctly in favour of evolution. Indeed I may go further. I may say that the hypothesis of evolution explains the facts of Miocene, Pliocene, and Recent distribution, and that no other supposition even pretends to account for them. It is, indeed, a conceivable supposition that every species of Rhinoceros and every species of Hyæna, in the long succession of forms between the Miocene and the present species, was separately constructed out of dust, or out of nothing, by supernatural power; but until I receive distinct evidence of the fact, I refuse to run the risk of insulting any sane man by supposing that he seriously holds such a notion.

Let us now take a step further back in time, and inquire into the relations between the Miocene Fauna and its predecessor of the Upper Eocene formation.

Here it is to be regretted that our materials for forming a judgment are nothing to be compared in point of extent or variety with those which are yielded by the Miocene strata. However, what we do know of this Upper Eocene Fauna of Europe gives sufficient positive information to enable us to draw some tolerably safe inferences. It has yielded representatives of *Insectivora*, of *Cheiroptera*, of *Rodentia*, of *Carnivora*, of artiodactyle and perissodactyle *Ungulata*, and of opossum-like Marsupials. No Australian type of Marsupial has been discovered in the Upper Eocene strata, nor any Edentate mammal. The genera (except perhaps in the case of some of the *Insectivora*, *Cheiroptera*, and *Rodentia*) are different from those of the Miocene epoch, but present a remarkable general similarity to the Miocene and recent genera. In several cases, as I have already shown, it has now been clearly made out that the relation between the Eocene and Miocene forms is such that the Eocene form is the less specialised; while its Miocene ally is more so, and the specialisation reaches its maximum in the recent forms of the same type.

So far as the Upper Eocene and the Miocene Mammalian Faunæ are comparable, their relations are such as in no way to oppose the hypothesis that the older are the progenitors of the more recent forms, while, in some cases, they distinctly favour that hypothesis. The period in time and the changes in physical geography represented by the nummulitic deposits are undoubtedly very

great, while the remains of Middle Eocene and
Older Eocene Mammals are comparatively few.
The general facies of the Middle Eocene Fauna,
however, is quite that of the Upper. The Older
Eocene pre-nummulitic mammalian Fauna contains Bats, two genera of *Carnivora*, three genera
of *Ungulata* (probably all perissodactyle), and a
didelphid Marsupial; all these forms, except
perhaps the Bat and the Opossum, belong to
genera which are not known to occur out of the
Lower Eocene formation. The *Coryphodon* appears
to have been allied to the Miocene and later
Tapirs, while *Pliolophus*, in its skull and dentition,
curiously partakes of both artiodactyle and perissodactyle characters; the third trochanter upon
its femur, and its three-toed hind foot, however,
appear definitely to fix its position in the latter
division.

There is nothing, then, in what is known of the
older Eocene mammals of the Arctogæal province
to forbid the supposition that they stood in an
ancestral relation to those of the Calcaire Grossier
and the Gypsum of the Paris basin, and that our
present fauna, therefore, is directly derived from
that which already existed in Arctogæa at the
commencement of the Tertiary period. But if
we now cross the frontier between the Cainozoic
and the Mesozoic faunæ, as they are preserved
within the Arctogæal area, we meet with an
astounding change, and what appears to be a

complete and unmistakable break in the line of biological continuity.

Among the twelve or fourteen species of *Mammalia* which are said to have been found in the Purbecks, not one is a member of the orders *Cheiroptera, Rodentia, Ungulata,* or *Carnivora,* which are so well represented in the Tertiaries. No *Insectivora* are certainly known, nor any opossum-like Marsupials. Thus there is a vast negative difference between the Cainozoic and the Mesozoic mammalian faunæ of Europe. But there is a still more important positive difference, inasmuch as all these Mammalia appear to be Marsupials belonging to Australian groups, and thus appertaining to a different distributional province from the Eocene and Miocene marsupials, which are Austro-Columbian. So far as the imperfect materials which exist enable a judgment to be formed, the same law appears to have held good for all the earlier Mesozoic *Mammalia.* Of the Stonesfield slate mammals, one, *Amphitherium,* has a definitely Australian character; one, *Phascolotherium,* may be either Dasyurid or Didelphine; of a third, *Stereognathus,* nothing can at present be said. The two mammals of the Trias, also, appear to belong to Australian groups.

Every one is aware of the many curious points of resemblance between the marine fauna of the European Mesozoic rocks and that which now

exists in Australia. But if there was this
Australian facies about both the terrestrial and
the marine faunæ of Mesozoic Europe, and if
there is this unaccountable and immense break
between the fauna of Mesozoic and that of
Tertiary Europe, is it not a very obvious suggestion that, in the Mesozoic epoch, the Australian
province included Europe, and that the Arctogæal
province was contained within other limits ? The
Arctogæal province is at present enormous, while
the Australian is relatively small. Why should
not these proportions have been different during
the Mesozoic epoch ?

Thus I am led to think that by far the simplest
and most rational mode of accounting for the
great change which took place in the living
inhabitants of the European area at the end of
the Mesozoic epoch, is the supposition that it
arose from a vast alteration of the physical
geography of the globe ; whereby an area long
tenanted by Cainozoic forms was brought into
such relations with the European area that
migration from the one to the other became
possible, and took place on a great scale.

This supposition relieves us, at once, from the
difficulty in which we were left, some time ago,
by the arguments which I used to demonstrate
the necessity of the existence of all the great
types of the Eocene epoch in some antecedent
period.

It is this Mesozoic continent (which may well have lain in the neighbourhood of what are now the shores of the North Pacific Ocean) which I suppose to have been occupied by the Mesozoic *Monodelphia ;* and it is in this region that I conceive they must have gone through the long series of changes by which they were specialised into the forms which we refer to different orders. I think it very probable that what is now South America may have received the characteristic elements of its mammalian fauna during the Mesozoic epoch; and there can be little doubt that the general nature of the change which took place at the end of the Mesozoic epoch in Europe was the upheaval of the eastern and northern regions of the Mesozoic sea-bottom into a westward extension of the Mesozoic continent, over which the mammalian fauna, by which it was already peopled, gradually spread. This invasion of the land was prefaced by a previous invasion of the Cretaceous sea by modern forms of mollusca and fish.

It is easy to imagine how an analogous change might come about in the existing world. There is, at present, a great difference between the fauna of the Polynesian Islands and that of the west coast of America. The animals which are leaving their spoils in the deposits now forming in these localities are widely different. Hence, if a gradual shifting of the deep sea, which at present bars

migration between the easternmost of these islands
and America, took place to the westward, while
the American side of the sea-bottom was gradually
upheaved, the palæontologist of the future would
find, over the Pacific area, exactly such a change
as I am supposing to have occurred in the North-
Atlantic area at the close of the Mesozoic period.
An Australian fauna would be found underlying
an American fauna, and the transition from the
one to the other would be as abrupt as that
between the Chalk and lower Tertiaries; and as
the drainage-area of the newly formed extension
of the American continent gave rise to rivers and
lakes, the mammals mired in their mud would
differ from those of like deposits on the Australian
side, just as the Eocene mammals differ from those
of the Purbecks.

How do similar reasonings apply to the other
great change of life—that which took place at the
end of the Palæozoic period?

In the Triassic epoch, the distribution of the
dry land and of terrestrial vertebrate life appears
to have been, generally, similar to that which
existed in the Mesozoic epoch; so that the Triassic
continents and their faunæ seem to be related to the
Mesozoic lands and their faunæ, just as those of the
Miocene epoch are related to those of the present
day. In fact, as I have recently endeavoured to
prove to the Society, there was an Arctogæal continent and an Arctogæal province of distribution
in Triassic times as there is now; and the *Saurop-*

sida and *Marsupialia* which constituted that fauna were, I doubt not, the progenitors of the *Sauropsida* and *Marsupialia* of the whole Mesozoic epoch.

Looking at the present terrestrial fauna of Australia, it appears to me to be very probable that it is essentially a remnant of the fauna of the Triassic, or even of an earlier, age ; [1] in which case Australia must at that time have been in continuity with the Arctogæal continent.

But now comes the further inquiry, Where was the highly differentiated Sauropsidan fauna of the Trias in Palæozoic times ? The supposition that the Dinosaurian, Crocodilian, Dicynodontian, and Plesiosaurian types were suddenly created at the end of the Permian epoch may be dismissed, without further consideration, as a monstrous and unwarranted assumption. The supposition that all these types were rapidly differentiated out of *Lacertilia* in the time represented by the passage from the Palæozoic to the Mesozoic formation, appears to me to be hardly more credible, to say nothing of the indications of the existence of Dinosaurian forms in the Permian rocks which have already been obtained.

For my part, I entertain no sort of doubt that the Reptiles, Birds, and Mammals of the Trias are the direct descendants of Reptiles, Birds, and Mammals which existed in the latter part of the

[1] Since this Address was read, Mr. Krefft has sent us news of the discovery in Australia of a freshwater fish of strangely Palæozoic aspect, and apparently a Ganoid intermediate between *Dipterus* and *Lepidosiren*. [The now well-known *Ceratodus*. 1894.]

Palæozoic epoch, but not in any area of the present dry land which has yet been explored by the geologist.

This may seem a bold assumption, but it will not appear unwarrantable to those who reflect upon the very small extent of the earth's surface which has hitherto exhibited the remains of the great Mammalian fauna of the Eocene times. In this respect, the Permian land Vertebrate fauna appears to me to be related to the Triassic much as the Eocene is to the Miocene. Terrestrial reptiles have been found in Permian rocks only in three localities; in some spots of France, and recently of England, and over a more extensive area in Germany. Who can suppose that the few fossils yet found in these regions give any sufficient representation of the Permian fauna?

It may be said that the Carboniferous formations demonstrate the existence of a vast extent of dry land in the present dry-land area, and that the supposed terrestrial Palæozoic Vertebrate Fauna ought to have left its remains in the Coal-measures, especially as there is now reason to believe that much of the coal was formed by the accumulation of spores and sporangia on dry land. But if we consider the matter more closely, I think that this apparent objection loses its force. It is clear that, during the Carboniferous epoch, the vast area of land which is now covered by Coal-measures must have been undergoing a gradual depression. The dry land thus depressed

must, therefore, have existed, as such, before the
Carboniferous epoch—in other words, in Devonian
times—and its terrestrial population may never
have been other than such as existed during the
Devonian, or some previous epoch, although much
higher forms may have been developed elsewhere.

Again, let me say that I am making no
gratuitous assumption of inconceivable changes.
It is clear that the enormous area of Polynesia is,
on the whole, an area over which depression has
taken place to an immense extent; consequently
a great continent, or assemblage of subcontinental
masses of land must have existed at some former
time, and that at a recent period, geologically
speaking, in the area of the Pacific. But if that
continent had contained Mammals, some of them
must have remained to tell the tale; and as it is
well known that these islands have no indigenous
Mammalia, it is safe to assume that none existed.
Thus, midway between Australia and South
America, each of which possesses an abundant
and diversified mammalian fauna, a mass of land,
which may have been as large as both put together,
must have existed without a mammalian in-
habitant. Suppose that the shores of this great
land were fringed, as those of tropical Australia are
now, with belts of mangroves, which would extend
landwards on the one side, and be buried beneath
littoral deposits on the other side, as depression
went on; and great beds of mangrove lignite

might accumulate over the sinking land. Let upheaval of the whole now take place, in such a manner as to bring the emerging land into continuity with the South-American or Australian continent, and, in course of time, it would be peopled by an extension of the fauna of one of these two regions—just as I imagine the European Permian dry land to have been peopled.

I see nothing whatever against the supposition that distributional provinces of terrestrial life existed in the Devonian epoch, inasmuch as M. Barrande has proved that they existed much earlier. I am aware of no reason for doubting that, as regards the grades of terrestrial life contained in them, one of these may have been related to another as New Zealand is to Australia, or as Australia is to India, at the present day. Analogy seems to me to be rather in favour of, than against, the supposition that while only Ganoid fishes inhabited the fresh waters of our Devonian land, *Amphibia* and *Reptilia*, or even higher forms, may have existed, though we have not yet found them. The earliest Carboniferous *Amphibia* now known, such as *Anthracosaurus*, are so highly specialised that I can by no means conceive that they have been developed out of piscine forms in the interval between the Devonian and the Carboniferous periods, considerable as that is. And I take refuge in one of two alternatives : either they existed in our own area during the Devonian epoch and we have simply not yet found them; or they formed part of the population of some other distributional province of that day, and only entered our area by migration at the end of the Devonian epoch. Whether *Reptilia* and *Mammalia* existed along with them is to me, at present, a perfectly open question, which is just as likely to receive an affirmative as a negative answer from future inquirers.

Let me now gather together the threads of my argumentation into the form of a connected hypothetical view of the manner in which the distribution of living and extinct animals has been brought about.

I conceive that distinct provinces of the distribution of terrestrial life have existed since the earliest period at which that life is recorded, and possibly much earlier; and I suppose, with Mr. Darwin, that the progress of modification of terrestrial forms is more rapid in areas of elevation than in areas of depression. I take it to be certain that Labyrinthodont *Amphibia* existed in the distributional province which included the dry land depressed during the Carboniferous epoch; and I conceive that, in some other distributional provinces of that day, which remained in the condition of stationary or of increasing dry land, the various types of the terrestrial *Sauropsida* and of the *Mammalia* were gradually developing.

The Permian epoch marks the commencement of a new movement of upheaval in our area, which attained its maximum in the Triassic epoch, when dry land existed in North America, Europe, Asia, and Africa, as it does now. Into this great new continental area the Mammals, Birds, and Reptiles developed during the Palæozoic epoch spread, and formed the great Triassic Arctogæal province. But, at the end of the Triassic period, the movement of depression recommenced in our area, though it was doubtless balanced by elevation elsewhere; modification and development, checked in the one province, went on in that "elsewhere"; and the chief forms of Mammals, Birds and Reptiles, as we know them, were evolved and peopled the Mesozoic continent. I conceive Australia to have become separated from the continent as early as the end of the Triassic epoch, or not much later. The Mesozoic continent must, I conceive, have lain to the east, about the shores of the North Pacific and Indian Oceans; and I am inclined to believe that it continued along the eastern side of the Pacific area to what is now the province of Austro-Columbia, the characteristic fauna of which is probably a remnant of the population of the latter part of this period.

Towards the latter part of the Mesozoic period the movement of upheaval around the shores of the Atlantic once more recommenced, and was very probably accompanied by a depression around those of the Pacific. The Vertebrate fauna elaborated in the Mesozoic continent moved westward and took possession of the new lands, which gradually increased in extent up to, and in some directions after, the Miocene epoch.

It is in favour of this hypothesis, I think, that it is consistent with the persistence of a general uniformity in the positions of the great masses of land and water. From the Devonian period, or earlier, to the present day, the four great oceans, Atlantic, Pacific, Arctic, and Antarctic, may have occupied their present positions, and only their coasts and channels of communication have undergone an incessant alteration. And, finally, the hypothesis I have put before you requires no supposition that the rate of change in organic life has been either greater or less in ancient times than it is now ; nor any assumption, either physical or biological, which has not its justification in analogous phenomena of existing nature.

I have now only to discharge the last duty of my office, which is to thank you, not only for the patient attention with which you have listened to me so long to-day, but also for the uniform kindness with which, for the past two years, you have rendered my endeavours to perform the important, and often laborious, functions of your President a pleasure instead of a burden.

END OF VOLUME VIII.

Milton Keynes UK
Ingram Content Group UK Ltd.
UKHW032319161024
449665UK00001B/42

9 781108 040587